CISM COURSES AND LECTURES

Series Editors:

The Rectors of CISM
Sandor Kaliszky - Budapest
Mahir Sayir - Zurich
Wilhelm Schneider - Wien

The Secretary General of CISM
Giovanni Bianchi - Milan

Executive Editor
Carlo Tasso - Udine

The series presents lecture notes, monographs, edited works and proceedings in the field of Mechanics, Engineering, Computer Science and Applied Mathematics.
Purpose of the series is to make known in the international scientific and technical community results obtained in some of the activities organized by CISM, the International Centre for Mechanical Sciences.

INTERNATIONAL CENTRE FOR MECHANICAL SCIENCES

COURSES AND LECTURES - No. 382

LEARNING, NETWORKS AND STATISTICS

EDITED BY

G. DELLA RICCIA
UNIVERSITY OF UDINE

H.-J. LENZ
FREE UNIVERSITY OF BERLIN

R. KRUSE
UNIVERSITY OF MAGDEBURG

  Springer-Verlag Wien GmbH

Le spese di stampa di questo volume sono in parte coperte da
contributi del Consiglio Nazionale delle Ricerche.

This volume contains 53 illustrations

In order to make this volume available as economically and as
rapidly as possible the authors' typescripts have been
reproduced in their original forms. This method unfortunately
has its typographical limitations but it is hoped that they in no
way distract the reader.

ISBN 978-3-211-82910-3   ISBN 978-3-7091-2668-4 (eBook)
DOI 10.1007/978-3-7091-2668-4

PREFACE

This volume contains the papers accepted for presentation at the invitational ISSEK96 workshop on 'Learning, Networks and Statistics' organized by the International School for the Synthesis of Expert Knowledge (ISSEK) and held at the Centre International des Sciences Mécaniques (CISM) in Udine from 19 to 21 September, 1996.

The first workshop on "Mathematical and Statistical Methods in Artificial Intelligence" organised by the International School for the Synthesis of Expert Knowledge (ISSEK) was held in 1994. This brought together a group of scientists with international reputation in the field of Mathematics, Statistics and Computer Science and it was a great success since all the participants had a stong interest in interdisciplinary work and collaboration. Instead of a call for papers the participants were recruited according to their scientific standings and by individual invitation. For further information on this event we refer to the book: Mathematical and Statistical Methods in Artificial Intelligence, G. Della Riccia, R. Kruse and R. Viertl (eds.); CISM Courses and Lecture No. 363, International Centre for Mechanical Sciences (Spriger-Verlag Wien New York).

In a joint venture of the Free University of Berlin (Prof. Lenz), the Otto-von-Guericke University of Magdeburg (Prof. Kruse) and the University of Udine (Prof. Della Riccia), a second international workshop on 'Learning, Networks and Statistics' was held. According to the overwhelming success of the first workshop the same fundamental principles were followed up, in particular, the organisers invited a rather limited but distinguished group of speakers with superb methodological backgrounds and with an expressed interest in overcoming the traditional frontiers of Computer Science, Mathematics and Applied Statistics.

Motivated by a recent book published by D. Michie, D. J. Spiegelhalter and C. C. Taylor (Machine Learning, Neural and Statistical Classification, Ellis Horwood, 1994), the workshop was organised in 4 sessions reflecting the main areas of interest: Neural Nets, Statistics and Networks, Classification and Data Mining and (Machine) Learning.

As it was intended not to have a pure academic discussion, but to inspect and unwrap practical problems too, the organisers as well as the editors were glad to have a lot of stimulating contributions from the 'shop-floor'. The various fields of interest of the participants and the stimulating atmosphere of the lecture hall at Palazzo del Torso of CISM at Udine made it possible to analyse some important topics with stepwise refinement and on a uniform terminology basis. The editors believe that those discussions helped to clarify several of our ideas and helped to polish the papers accepted for final publication and collected in this volume.

The organisers and the editors of this volume would like to thank the following organisations which sponsored this event and made a meeting of such calibre possible:

-The International School for the Synthesis of Expert Knowledge (ISSEK) for promoting the workshop.
-The Free University of Berlin and the University of Udine for their administrative support.
- The Centre International des Sciences Mécaniques (CISM) for hosting again the workshop in their beautiful Palazzo del Torso and for providing technical support.

On behalf of all participants we express our gratitude to the Fondazione Cassa di Risparmio di Udine e Pordenone, to the Cassa di Risparmio di Udine and Pordenone and to EuroStat (Statistical Office of the European Communities at Luxembourg) for granting the necessary financial support.

Finally, we would like to thank very warmly Mrs. Angelika Wnuk from the Free University of Berlin for her superb secretarial work throughout the planning period and the time after.

The editors are happy and proud to announce a third event of this kind to be held in 1998 and are looking forward to meeting again most members of this scientific community.

H.-J. Lenz
R. Kruse
G. Della Riccia

CONTENTS

NEURAL NETS

OVERTRAINING IN SINGLE-LAYER PERCEPTRONS

S. Raudys
Vilnius Gediminas Technical University , Vilnius, Lithuania

ABSTRACT

The "overtraining" takes origin in different surfaces of learning-set and test-set cost functions. Essential peculiarities of non-linear single-layer perceptron training are the growth of magnitude of weights with an increase in number of iterations and, as a consequence, a change of the criterion used to find the weights. Therefore in non-linear SLP training one can obtain various statistical classification rules of different complexity and, the overtraining preventing problem can be analyzed as a problem of selecting the proper type of statistical classifier. Which classifier is the best one for a given situation depends on the number of features, data size and its configuration. In order to obtain a wider range of classifiers in non-linear SLP training, several new complexity control procedures are suggested.

Key words: *overtraining, optimization criterion, maximal margin, targets, scissors effect, regularization, anti-regularization, generalization, learning set size, dimensionality.*

1. INTRODUCTION.

The generalization error is one of the principal characteristics of any pattern classifying system. It is easy to design a system that classifies well the training patterns, however, it is difficult to design a system that classifies well patterns not participating in the training process. Therefore, the generalization of statistical and neural net classifiers has been extensively studied during the last three decades.

When applying standard statistical methods to find weights of the classification rules, one estimates the weight values in one shot after solving a certain system of linear equations, matrix inversion, etc. In adaptive training, one needs to minimize a certain cost function and obtains the solution in an iterative way, step by step. Usually, with an increase in the number of iterations the cost function diminishes gradually with certain local fluctuations. In the statistical analysis, indentification, artificial neural net analysis, it was noticed that the test-set error (generalization error), however, decreases at first, but very often, after a certain number of iterations, begins increasing. In artificial neural net literature this effect was called *overtraining*.

In neural network training, we observe the overtraining effect nearly always. Sometimes we do not notice this effect at all. The objective of the present paper is to explain the origin of overtraining, to reveal its peculiarities in single-layer perceptron training, to find the ways how to prevent this effect or to utilise it in the best way. In this paper, we will analyze linear and non-linear single-layer perceptron classifiers, only.

The single-layer perceptron is the simplest decision-making unit, often called an elementary neuron. A group of elementary neurons organized in layers composes a multilayer perceptron. If used as a classifier, the single-layer perceptron (SLP) forms a linear decision boundary. The SLP has a number of inputs (say p) x_1, x_2, ... , x_p and one output calculated according to the following equation[1]

$$output = o(\mathbf{w}'\,\mathbf{x} + w_o), \tag{1}$$

where
$\mathbf{x}$ is a p-variate vector to be classified,
$o(net$) is a non-linear activation (transfer) function, e.g.,

$$o(net)= tanh(net)= (e^{net} - e^{-net}) / (e^{net} + e^{-net}), \tag{2}$$

w_o , $\mathbf{w}$ are weights of the discriminant function to be learned in the training process.

Usually the perceptron' weights are found in the iterative training procedure, where the following cost function of the sum of the squares

$$cost = \frac{1}{2(N_1 + N_2)} \sum_{i=1}^{2} \sum_{j=1}^{N_i} (t_j^{(i)} - o(\mathbf{w}'\mathbf{x}_j^{(i)} + w_o))^2 , \tag{3}$$

is minimized. In the above formula, $t_j^{(i)}$ is a desired output (a target) for $\mathbf{x}_j^{(i)}$, the j-th learning set observation vector from the i-th class, N_i is the number of learning vectors from the i-th class. For activation function (2), one usually uses $t_j^{(1)} = 1$ and $t_j^{(2)} = -1$, or $t_j^{(1)} = 0.9$ and $t_j^{(2)} = -0.9$.

Different optimization techniques can be used to minimize cost function (3). In the standard global gradient delta learning rule (back propagation - BP), the weight vector is adapted according to the following rule

$$\mathbf{w}_{(t+1)} = \mathbf{w}_{(t)} - \eta \frac{\partial cost_t}{\partial \mathbf{w}}, \tag{4}$$

where η is called a learning step.

If the activation function is linear, i.e., $o(net)= net$, then we have a linear perceptron, otherwise, the non-linear SLP. We see that use of the non-linear activation function introduces specific peculiarities in the non-linear SLP training process and the overtraining analysis.

The paper is organised as follows. In the second section, we analyze a linear SLP classifier and explain that overtraining arises due to different cost functions of the learning-set and the test-set. In the third section, when analyzing the non-linear SLP we show that in the batch-mode training, in the way between the start point and the minimum of the cost function, the cost function changes, and one obtains a sequence of statistical classifiers of different complexity. A link between the generalization error, complexity, and learning set size is considered in the fourth section. A "scissors effect" is explained which indicates that in the case of a small learning-set, it is preferable to use simple classifiers instead of complex ones. Thus, the overtraining preventing problem can be analyzed as a problem of selecting the proper type of statistical classifier. The fifth section contains simulation results showing that, depending on data size and its configuration one has to train the perceptron one or the maximal number of iterations. The means of controling the complexity are reviewed in the sixth section. Several new complexity control procedures are proposed. The seventh section presents conclusions.

2. OVERTRAINING IN THE LINEAR SLP.

In order to understand the overtraining effect one needs to take into account that there exist *two different cost functions:*

1) cost function *cost* to be minimized in the training process (this cost function is based on the learning-set pattern vectors) and
2) cost function C_t of the classification error measured on general population (in practice, it is measure on a separate test-set). This error is usually called *a conditional classification error or the generalization error.*

Each cost function has its own landscape, and a minimization of the first function does not imply the minimization of the second one. We'll explain this phenomenon graphically by the following example.

Example. Two landscapes depicted in Fig. 1 correspond to the simple linear adaptive classifier ADALINE trained by using the conventional cost function of the sum of squares (3) without non-linearity and targets +1 and -1. The data: two univariate spherical Gaussian distributions $N(0,1)$ and $N(4,1)$. In the univariate case we have only two independent parameters of the linear discriminant function $g(x_1) = w_1*x_1+w_0$. The landscape *cost* (dots) with the minimum at W^l ($w_0 = 0.56$, $w_1 = -0.37$) was evaluated on the training-set composed of 8 randomly chosen univariate observations in each class ($N_1 = N_2 = 8$) from the *cost function (3) represented as*

$$cost = 1 + \frac{1}{2}\mathbf{w}'(\mathbf{K}_1 + \mathbf{K}_2)\,\mathbf{w} + \frac{1}{2}w_o^2 + \mathbf{w}'(\overline{\mathbf{x}}_1 - \overline{\mathbf{x}}_2) + \mathbf{w}'(\overline{\mathbf{x}}_1 + \overline{\mathbf{x}}_2)w_o, \qquad (5)$$

where $\mathbf{K}_i = \dfrac{1}{N}\displaystyle\sum_{j=1}^{N} \mathbf{x}\,_j^{(i)}\,((\mathbf{x}\,_j^{(i)})'$ and $\overline{\mathbf{x}}_i$ is i-th class sample mean vector .

The landscape C_t (solid line) was calculated theoretically for the cost function evaluated for true class conditional densities

$$C_t = 1 + \frac{1}{2}\mathbf{w}'(\Sigma_1 + \Sigma_2 + \mu_1\,\mu_1' + \mu_2\mu_2')\,\mathbf{w} + \frac{1}{2}w_o^2 + \mathbf{w}'(\mu_1 - \mu_2) + \mathbf{w}'(\mu_1 + \mu_2)w_o, \qquad (6)$$

where Σ_i is the covariance matrix of i-th class and μ_i is its mean vector. In (5) and (6), we have used multivariate notations.

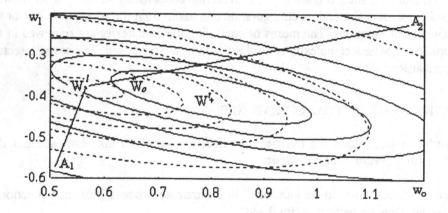

Fig. 1. Two cost function landscapes illustrating the overtraining effect. Training performed from A_2 to W^l will lead to overtraining and start from A_1 will not.

The landscape plotted in dotted lines has a minimum in W^l. The goal of training is to find the minimum of this function (point W^l). The landscape plotted in solid lines corresponds to the cost function evaluated from the test-set data (generalization error). Let us start the search for the minimum of the cost function *cost* with the point A_2. Suppose we succeed in finding an absolutely straight way from A_2 to W^l. Then the cost function C_t will decrease at first, reach the minimum at point W_o ($w_o = 0.65$, $w_1 = -0.35$), and aferwards on the way to the point W^l it will increase again. Clearly, minimization of $cost_t$ does not lead to W^t, the minimum of C_t. Instead, we arrive to W^l, the minimum of *cost*. Thus we have overtraining. However, if we started from the point A_1 and went straight to W^l, then we would not obtain overtraining.

Training performed starting from A_2 and going to W^l will lead to overtraining and the start from A_1 will not. Thus, the presence of overtraining depends on the starting point. In a random selection of the starting point, the overtraining effect is observed only in a part of training procedures. Thus we can explain the overtraining effect by *differences between the cost error surfaces obtained for the learning-set data and that obtained for the test-set data*. The larger the difference, the larger the overtraining effect can be expected. The difference depends on the number of training examples, the complexity of the algorithm, and also on individual peculiarities of random learning and test sets.

We see that in an adaptive training process, sometimes it is worth stoping earlier. A gain due to the early stopping is easily explained by the statistical analysis. Let us have a learning-set LS_0, of N_0 observations and estimate parameters of the network by the maximum likelihood method. Note that the minimization of cost function (3) with linear $o(net)$ is actually the maximum likelihood method. Then the sample estimate of the weight vector $\hat{w}_0$ obtained after minimizing the cost function is a random vector. It is asymptotically Gaussian: $\hat{w}_0 \sim N(w^*, 1/N_0 G^{-1})$, where w^* is an optimal weight vector obtained for the general population, and G is a Fisher information matrix (see, e.g., [2]).

Let us now have another independent learning-set LS_1, of N_1 observations and estimate parameters iteratively by minimizing cost function (3) starting from the weight vector $\hat{w}_0$ and gradually decreasing the learning-step value. When the learning process ends, we obtain a new estimate $\hat{w}_1$ of the weight. According to the statistical theory, asymptotically $\hat{w}_1 \sim N(w^*, 1/N_1 G^{-1})$. If $N_1 < N_0$, then, on average, we have a lower accuracy. Obviously, adaptive estimation of the weight vector did not allow us to utilise the information contained in the first learning-set LS_0. Let us now use the following estimate of the weight vector

$$\hat{w}_{new} = \alpha \, \hat{w}_0 + (1-\alpha) \, \hat{w}_1. \tag{7}$$

Asymptotically, for large sample sizes N_0 and N_1, the vector $\hat{w}_{new}$ will have a Gaussian distribution: $\hat{w}_{new} \sim N(w^*, (\alpha^2/N_1 + (1-\alpha)^2/N_2) \, G^{-1})$. Equating derivative of (7) with respect to α to zero and solving the linear equation indicate that smallest covariance of the weight $\hat{w}_{new}$ will be obtained, when

$$\alpha^{optimal \, s} = N_0 / (N_0 + N_1). \tag{8}$$

Then statistically optimal estimate $\hat{w}_{new}^{optimal \, s} \sim N(w^*, 1/(N_1+N_2)G^{-1})$. Therefore, in adaptive estimation of the weight vector, early stopping can increase the estimation accuracy. If one stops optimally, then one will obtain the same accuracy as in a joint use of both learning sets, LS_0 and LS_1. It is essentially the overtraining effect: while moving from $\hat{w}_0$ towards $\hat{w}_1$, the accuracy increases at first, and afterwards begins decreasing. Use of knowledge on small learning sample statistical properties of the estimates allows us to determine the optimal stopping moment (in our case, we use the weighted estimate(7)) and the optimal weight estimate.

Suppose, the matrix $G=I$ (identity matrix). Then a squared Euclidean distance $d_1^2 = (\hat{w}_1 - w^*)'(\hat{w}_1 - w^*)$ between the estimate $\hat{w}_1$ and the optimal weight vector w^* will be a chi-square scaled random variable with mean p/N_1 and standard deviation $\sqrt{2p}/N_1$. For large p, $(\hat{w}_1 - w^*)'(\hat{w}_1 - w^*) \approx p/N_1$, and for the statistically optimal weighted estimate $(\hat{w}_{new}^{otimal} - w^*)'(\hat{w}_{new}^{otimal} - w^*) \approx p/(N_0 + N_1)$. The above estimate shows that the largest gain due to the optimal early stopping, can be obtained when $N_0 \gg N_1$.

Next we demonstrate that utilization of additional information can increase the accuracy even more. Suppose, a trajectory of iterative move from $\hat{w}_0$ to $\hat{w}_1$ is linear. All the three points w^*, $\hat{w}_0$, $\hat{w}_1$ and the trajectory from point $\hat{w}_0$, to point $\hat{w}_1$ can be depicted on the plane (see Fig. 2). Suppose, $N_1 = 4N_0$, the matrix $G=I$ (identity matrix), a dimensionality is high and we analyze one learning set LS_0 and several learning sets LS_1^a, LS_1^b, LS_1^c, LS_1^d, LS_1^e. Then the we have one starting point $\hat{w}_0$ and several final points $\hat{w}_1^a$, $\hat{w}_1^b$, $\hat{w}_1^c$, $\hat{w}_1^d$, $\hat{w}_1^e$, situated on a p-dimensional hyper-sphere of radius d_1 and the centre w^*; all Euclidean distances $d_1 = d_0/2$.

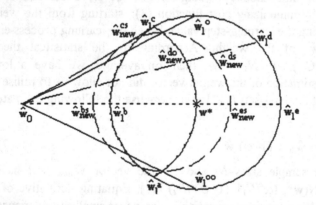

Fig.2. Use of early stopping to prevent overtraining when moving from the starting point $\hat{w}_0$ to $\hat{w}_1$.

According to the statistically-based estimate (8) for sample size $N_1 = 4N_0$ one has to stop at $\frac{4}{5}$ distance from $\hat{w}_0$ to $\hat{w}_1$. Statistically optimal stopping points compose a hyper-sphere of radius $\frac{4}{5}d_1$ (a dashed circle containing points $\hat{w}_{new}^{bs}$, $\hat{w}_{new}^{cs}$, $\hat{w}_{new}^{ds}$ in Fig. 2). Fig. 2, however, indicates that for the learning sets LS_1^a, LS_1^b, and LS_1^c it is better to stop at the very end of training, i.e. at $\hat{w}_1^a$, $\hat{w}_1^b$, and $\hat{w}_1^c$. For the learning sets LS_1^d and LS_1^e, it is better to stop much earlier, i.e., at $\hat{w}_{new}^{do}$ and $\hat{w}_{new}^{eo} = w^*$. It is a typical situation in the small-dimensional case. In order to stop optimally, for each individual pair of points $\hat{w}_0$ and $\hat{w}_1^*$, one needs additional information never used in training the perceptron.

In the high-dimensional case, however, the standard deviation $\sqrt{2p}/N_1 \ll p/N_1$, the mean. The distant and close points, such as $\hat{w}_1^c$, $\hat{w}_1^b$, will be very rare. Thus, all the distances $(\hat{w}_1-\hat{w}_0)'(\hat{w}_1-\hat{w}_0) \approx p/N_0 + p/N_1$ (see points $\hat{w}_1^o$ and $\hat{w}_1^{oo}$ in Fig. 2) and the statistical estimate (8) will be close to the optimal one for all learning-sets.

Apparently, assumptions on probabilistic structure of the data and consequent theoretical considerations can help to estimate the average value of the optimal stopping moment. Emplyment of the additional validation set is a universal method. It is, in fact, a special way to extract the useful information contained in the validation set.

3. SEVEN STATISTICAL CLASSIFIERS IN THE NON-LINEAR SLP DESIGN

Overtraining in the linear SLP is caused by differences in the cost function surfaces determined by the learning-set and by the general population. In the classification problem, however, one extra factor affects the difference in the cost functions: during the learning phase one minimizes the cost function of the sum of squares, and actually one is interested in the probability of misclassification (the generalization error).

Non-linearity of the activation function gives rise *to one more factor* in differences of the cost functions mentioned. During the iterative training process, in the way between the starting point and the minimum of the cost function, the SLP can realize several statistical classifiers of different complexity [3]. Different finite learning sample properties of these classifiers pose a problem of selecting one of them that is the best for this particular situation. Below, we present some details.

Let $t_j^{(1)} = 1$ and $t_j^{(2)} = -1$, $\overline{x}^{(1)} + \overline{x}^{(2)} = 0$, $N_2 = N_1$, the prior weights are zero, i.e., $w_{0(0)} = 0$, $\mathbf{w}_{(0)} = 0$, and we train the SLP in the batch-mode. Therefore during the first iterations, while the weights and a scalar product, $w_o + \mathbf{w}'\mathbf{x}_j^{(i)}$ are close to zero, $\partial\, o(v) = dv$, and $o(w_o + \mathbf{w}'\mathbf{x}_j^{(i)}) = w_o + \mathbf{w}'\mathbf{x}_j^{(i)}$. Then

$$\frac{\partial\, cost}{\partial w_o}\bigg|_{w_o = w_{o(t)}} = -\frac{1}{2N}\sum_{i=1}^{2}\sum_{j=1}^{N}(t_j^{(i)} - w_{o(t)} - (\mathbf{x}_j^{(i)})'\mathbf{w}_{(t)}) = w_{o(t)} + \frac{1}{2}(\overline{x}^{(1)} + \overline{x}^{(2)})'\mathbf{w}_{(t)} = 0,$$

$$\frac{\partial\, cost_1}{\partial w}\bigg|_{w = w_{(t)}} = -\frac{1}{2N}\sum_{i=1}^{2}\sum_{j=1}^{N}\mathbf{x}_j^{(i)}(t_j^{(i)} - w_{o(t)} - (\mathbf{x}_j^{(i)})'\mathbf{w}_{(t)}) = -\frac{1}{2}\Delta\overline{x} + \mathbf{K}\mathbf{w}_{(t)},$$

where $\Delta\overline{x} = \overline{x}^{(1)} - \overline{x}^{(2)}$, $\mathbf{K} = \frac{1}{2N}\sum_{i=1}^{2}\sum_{j=1}^{N}\mathbf{x}_j^{(i)}(\mathbf{x}_j^{(i)})' = \frac{N-1}{N}S + \frac{1}{4}\Delta\overline{x}\,\Delta\overline{x}',$

$\bar{x}^{(i)} = \dfrac{1}{N_i} \sum\limits_{j=1}^{N_i} x_j^{(i)}$ is a sample mean vector and

$$S = \dfrac{1}{N_1 + N_2 - 2} \sum\limits_{i=1}^{2} \sum\limits_{j=1}^{N_i} (x_j^{(i)} - \bar{x}^{(i)})(x_j^{(i)} - \bar{x}^{(i)})'$$

is a sample pooled covariance matrix. We regard the matrix K to be not singular and have its inversion.

After the first learning iteration, we get

$$w_{(1)} = \eta\, t_1\, k_1\, \Delta\bar{x}, \; w_{0(1)} = 0. \tag{9}$$

It is a weight vector of the Euclidean distance classifier [4]. It can be shown that after t iterations $w_{0(t)} = 0$, $w_{(t)} = \left(I - (I - \eta K)^t \right) \dfrac{1}{2} K^{-1} \Delta\bar{x}$,

Employing the first terms of expansions $(I - \eta K)^t = I - t\eta K + \dfrac{1}{2} t(t-1)\eta^2 K^2 - ...)$ and $(I - \beta K)^{-1} = I + \beta K + ...$ for small η and t and assuming the matrix $I\lambda + S$ to be not singular, after some matrix algebra we obtain

$$w_{(t)} = \left(I \dfrac{2}{(t-1)\eta} \dfrac{N}{N-1} + S \right)^{-1} \Delta\bar{x}\, k_R, \tag{10}$$

where

$$k_R = \dfrac{2\eta t}{(D_R^2 + 2\eta(t-1)(N-1)/N)}, \quad D_R^2 = \Delta\bar{x}' S_R^{-1} \Delta\bar{x}, \quad S_R = S + I\dfrac{2}{(t-1)\eta}\dfrac{N}{N-1}.$$

The weight vector $w_{(t)}$ is equivalent to that of *the regularized discriminant analysis* (for a general guide for the discriminant analysis problems see [5]) with the regularization parameter $\lambda = \dfrac{2}{(t-1)\eta}\dfrac{N}{N-1}$. Expression (10) indicates, that if the number of iterations is increasing, then $(I\dfrac{2}{(t-1)\eta}\dfrac{N}{N-1} + S) \to S$, $k_R \to 1$, and *the resulting classifier approaches the Fisher linear DF.*

In [3, 6] the expectation of the cost function $cost(t_i, X_j^{(i)})$ determined by Equation (3) with respect to random targets and the set of $2N$ random Gaussian $N(X, \mu_i, I)$ learning vectors $X_1^{(1)}, ..., X_N^{(2)}$ were analyzed. It has been shown that, in spite of non-linearity of activation function (2), the minimum of cost function (3) is obtained at

$$w^{opt} = (\mu_1 - \mu_2) \text{ and } w_o^{opt} = -\dfrac{1}{2}(\mu_1 + \mu_2)'w.$$

It means the magnitude of optimal weights is: $|w'\, w|^{\frac{1}{2}} = \delta$, where $\delta^2 = (\mu_1 - \mu_2)'(\mu_1 - \mu_2)$ is the squared Mahalanobis distance between the two pattern classes (recall that the

Mahalanobis distance δ and the asymptotic classification error P_∞ are related: $P_\infty = \Phi(-\delta/2)$). Therefore, for Gaussian patterns, during the training, the weights increase to reach certain values. These values depend on the separability of pattern classes δ. If the classes are well separated, then by minimizing (3) we have large weights.

Suppose that, after a certain number of iterations, the number of misclassifications becomes zero. Then one of the ways of reducing the cost function of the sum of squares (3) is to increase the values $net = \mathbf{w}'\mathbf{x}_j^{(1)} + w_o$ of training vectors in the first class, and to diminish (up to minus infinity) all the values $net = \mathbf{w}'\mathbf{x}_j^{(2)} + w_o$ of training vectors in the second class.

This can be done by increasing the magnitude of weights, i.e., as $|\mathbf{w}'\mathbf{w}|^{\frac{1}{2}} \to \infty$. Then $|net| \to \infty$, and the cost function will diminish up to zero. It is additional evidence that one can obtain large weights for well separated classes with the non-linear activation function.

The latter observation indicates that for small N, when $2N < p + 2$, and/or when the number of misclassification becomes zero, after proper training, *the weights can become very large*.

For large weights, the values of the function $net = \mathbf{w}'\mathbf{x}_j^{(i)} + w_o$ will be either very high and positive (for the first class vectors) or very high and negative (for the second class vectors). Therefore the learning-set pattern vectors yield the output of perceptrons either +1 or -1. So for all the learning-set vectors, *activation function (2) will act as a hard-limiting threshold function*. It means that, if one uses the global minimization techniques that enable us to avoid local minima, *then one will obtain a classifier similar to the minimum empirical error one*. It has been demonstrated [3] that when $2N < p$ and the weights are still small, criterion (3) leads to the Fisher linear DF with a pseudo-inversion of the covariance matrix. In the way from the Fisher classifier to the minimum empirical error one, the activation function is a smooth non-linear function. Consequently, the contribution of distant atypical observations to the cost function is smaller than in the case of the linear activation function. Hence the classifier is more robust to outlyers (atypical observations) and we have a classifier similar to the *generalized discriminant analysis* proposed by Randles *et. al.* [7].

Let the empirical error be zero, and let $D(\mathbf{x}^*, \mathbf{w})$ be the Euclidean distance between the discriminant hyper-plane $g(\mathbf{x}) = \mathbf{w}'\mathbf{x} + w_o = 0$ and the learning set vector $\mathbf{x}^*$ closest to it. Let $D(\mathbf{x}_+^+, \mathbf{w})$ be the Euclidean distance between the discriminant hyper-plane and the second learning set vector $\mathbf{x}_+^+$, closest to it and different from the vector $\mathbf{x}^*$. Then for activation function (2) and the targets $t_1 = 1$ and $t_2 = -1$ asymptotically, with an increase in

the magnitude of the weights $\|\mathbf{w}\|$, the ratio $\dfrac{(t_+^+ - o(\mathbf{w}'\mathbf{x}_+^+))^2}{(t_*^* - o(\mathbf{w}'\mathbf{x}_*^*))^2}$ will diminish up to zero. It means, a relative contribution of the second learning set vector $\mathbf{x}_+^+$ closest to the decision hyper-plane, will become insignificant. Thus, the learning algorithm will tend to move the decision hyper-plane farther from the closest learning set vector $\mathbf{x}^*$. When the learning process is over, several vectors $\{\mathbf{x}^*\}$ will be at the same distance from the discriminant hyper-plane $g(\mathbf{x}) = \mathbf{w}''\mathbf{x} + w_0 = 0$. Only the learning set vectors $\{\mathbf{x}^{(*)}\}$ closest to the discriminant hyper-plane (Cortes and Vapnik [8] call them supporting patterns) will contribute to the value of the cost function and to the final determination of the hyper-plane' position. As a result we obtain *a maximum margin classifier*.

We conclude that the adaptive character of finding the weights of the SLP can lead to seven different statistical classifiers (see the learning scheme in Fig. 3) :

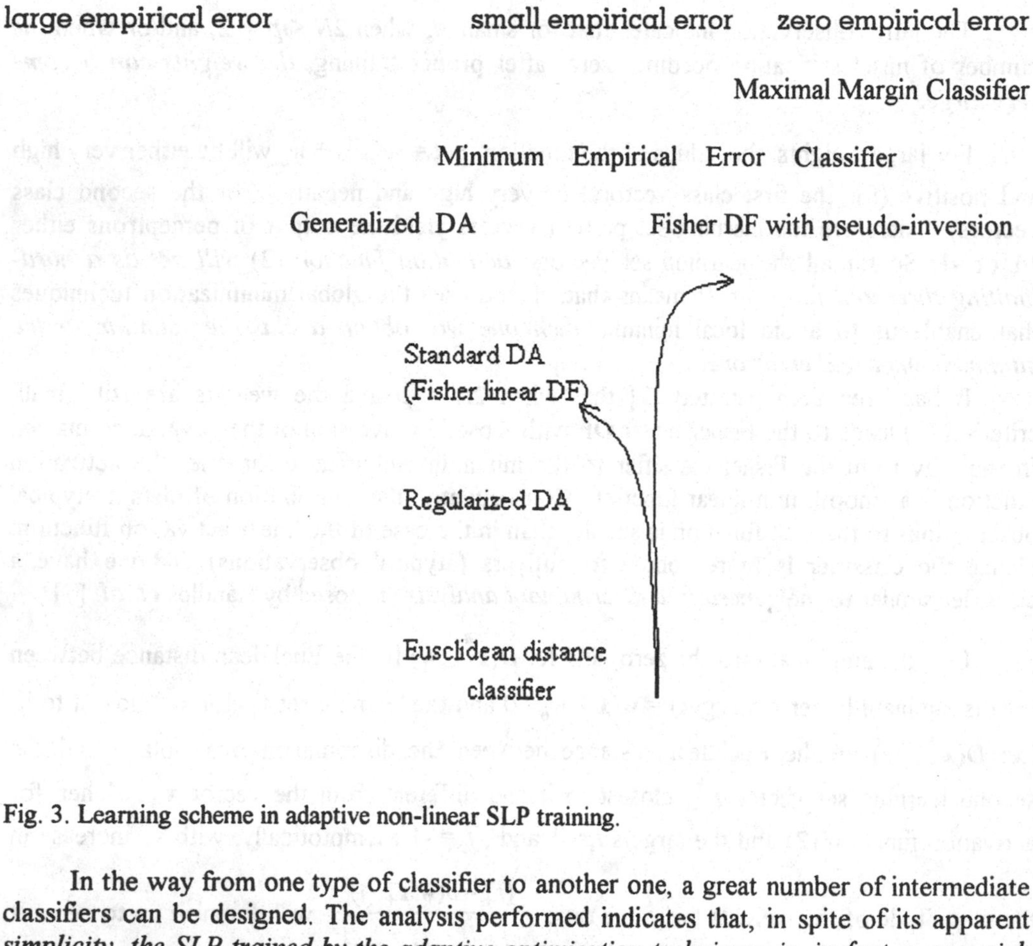

large empirical error   small empirical error   zero empirical error

Maximal Margin Classifier

Minimum   Empirical   Error   Classifier

Generalized DA   Fisher DF with pseudo-inversion

Standard DA
(Fisher linear DF)

Regularized DA

Eusclidean distance
classifier

Fig. 3. Learning scheme in adaptive non-linear SLP training.

In the way from one type of classifier to another one, a great number of intermediate classifiers can be designed. The analysis performed indicates that, in spite of its apparent simplicity, *the SLP trained by the adaptive optimization techniques is, in fact, a very rich family of linear classifiers.* We can assume that in principle, more types of the known classifiers can be obtained. Possibly, there exists a close link of multi-layer perceptrons with statistical techniques. There exists no distinct single-layer perceptron classifier. There is a great number of classifiers that can be obtained in training. In adaptive SLP training we deal with a continuum of statistical classifiers.

   The means of controling the type of the classifier are reviewed in the sixth section.

4. COMPLEXITY OF THE STATISTICAL CLASSIFIERS AND SAMPLE SIZE CONSIDERATIONS

   Weights of the classification rules depend on the learning-set vectors. Thus, they are estimated inexactly and can be considered as random variables. The *classification error* P_N *of the sample based classification rule is a random variable, too*. It is called a conditional classification error. The expectation EP_N of the conditional classification error over all possible learning sets of sizes N_1, N_2 is called *an expected error (generalization error)*. The limit of EP_N when the learning set sizes N_1, N_2 are increasing without bounds is called *an asymptotic error* and denoted by P_∞.

   In this section we analyze a link between the generalization error, complexity, and learning set size. We explain the "scissors effect" which indicates that in the case of a small learning-set, it is preferable to use simple classifiers instead of complex ones. Thus, the overtraining preventing problem can be analyzed as a problem of selecting the proper type of statistical classifier.

   The expected error of the Euclidean distance classifier for large N and p asymptotically is expressed in [9]

$$EP_N^E \approx \Phi\{ -\frac{\delta^*}{2} \frac{1}{\sqrt{1+\frac{2p^*}{N\delta^{*2}}}} \}, \tag{11}$$

where $\delta^* = \frac{\mu'\mu}{\sqrt{\mu'\Sigma\mu}}$ is a modified Mahalanobis distance, $p^* = \frac{(\mu'\mu)^2(tr\,\Sigma^2)}{(\mu'\Sigma\mu)^2}$ is a *modified dimensionality*, $\mu_1-\mu_2 = \mu$. For spherical Gaussian classes we have $\Sigma=I\sigma^2$, $\delta^{*2} = \delta^2 = \mu'\Sigma^{-1}\mu$, the squared Mahalanobis distance between the pattern classes, and $p^* = p$. If $\Sigma \neq I\sigma^2$, then $\delta^{*2} \leq \delta^2$ and $1 \leq p^* \leq \infty$. It means that, in theory, there exist situations where the Euclidean distance classifier is either very insensitive to the number of training observations or, on the contrary, its performance is very sensitive to the learning set size.

   The expected error of the standard Fisher DF [10,11]

$$EP_N^{(F)} \approx \Phi\{ -\frac{\delta}{2} \frac{1}{\sqrt{T_\mu T_\Sigma}} \}, \tag{12}$$

In Equation (12) the term $T_\mu = 1 + \dfrac{2p}{N\delta^2}$ appears due to inexact sample estimation of

the mean vectors μ_1, μ_2 of the classes, and term $T_S = 1 + \dfrac{p}{2N-p}$ due to inexact estimation

of the covariance matrix Σ. The above expression indicates that in the large learning-set

case, as $N \to \infty$, $EP_N^{(F)} \to \Phi\{ -\dfrac{\delta}{2} \} = P_\infty^{(F)} \le \Phi\{ -\dfrac{\delta^*}{2} \} = P_\infty^{(E)}$. Then it is more preferable to
use the Fisher classifier.

In the small learning-set case, the parameters of the classification rule are estimated
inexactly and the simple-structured Euclidean distance classifier can outperform the Fisher
classifier. It is a *"scissors effect"* known in Statistical pattern recognition for 25 years [12–
15]: in the small learning set cases, the use of simple structured classification rules is often
preferable to complex ones, and vice versa, in the large learning set cases, the complex
classifiers can be used more efficiently. An illustration of this phenomenon is presented in
Fig. 4, where, with a small learning-set size, Graph 1 (EDC) is significantly lower than
Graph 2 (Fisher classifier). Both graphs have been calculated using 20-variate Gaussian
data.

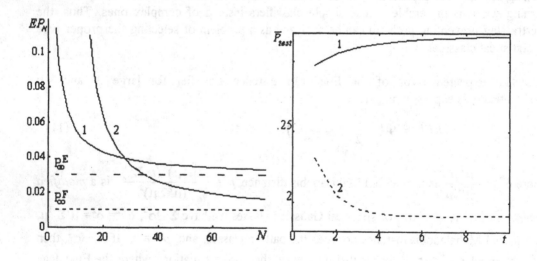

Fig. 4. The "scissors effect". The generalization
error versus N, the learning set size: 1 - EDC
for $p^*=20$, 2 - Fisher classifier. Graphs from [12].

Fig. 5. Generalization error as a function of
the number of learning iterations. 50-variate
Gaussian data; 1 - $N = 10$, 2 - $N = 500$.

In the context of SLP training it would mean that in the small learning-set case for
the correlated Gaussian data it is preferable to train the SLP only by one iteration.

Consequently, in the large learning-set case, it is preferable to train the perceptron more iterations. This conclusion is illustrated in Fig. 5, where we present the estimates of the generalization error $\overline{P}_t$, average values calculated over 10 independent repetitions of the experiment, as a function of the number of learning iterations t for two learning-set sizes: $N=10$ and $N=500$. In this simulation experiment we used 50-variate Gaussian data with unit variances and correlations between all the 50 variables $\rho = 0.05$. For this data $P_\infty^{(E)}=0.225$, $P_\infty^{(F)}=0.18$, $p^*=12$. In the SLP training we transfered the learning data centre $0.5(\overline{\mathbf{x}}^{(1)}+\overline{\mathbf{x}}^{(2)})$ into the zero point, initialized the SLP with the zero weight vector, used the sigmoid activation function and trained the perceptron in the batch-mode with the standard back propagation algorithm using the targets 0 and 1 and the learning step $\eta=5$.

For the regularized discriminant function an extra term appears in the asymptotic expression of the expected classification error [16]:

$$EP \approx \Phi\{-\frac{\delta^*}{2}\sqrt{\frac{1}{(1+\frac{2p}{N\delta^2})(1+\frac{2p}{2N-p})}}(1+\frac{\lambda\frac{p}{2N}}{1+\lambda\frac{2N}{2N-p}})\}. \tag{13}$$

In (13), the term $1+\lambda p/(2N(1+2\lambda N/(2N-p)))$ is apt to reduce a "negative" influence of estimation of the covariance matrix. After the optimization with respect to λ, the regularization parameter, the resulting generalization error is smaller than both the generalization error of the Euclidean distance and the Fisher classifier. In the previous section, we demonstrated that in the SLP training the number of learning steps plays the role of λ. Therefore the number of learning iterations can be used as a regularizer of the SLP classification rule.

In [17], the generalization error of the zero empirical error classifier and the margin classifier trained by a specific "random search" optimization procedure were analyzed for two equiprobable, spherical, Gaussian populations $N(\mu_i, I)$. This optimization procedure generates many random discriminant hyper-planes $\mathbf{w}'\mathbf{x}+w=0$ according to a certain *prior* distribution of the weights, determined by a priori density $f_{prior}(\mathbf{w}, w_o)$, and selects those that classify the learning sets LS^1 and LS^2, each of size N, without error. *A mean expected probability of misclassification* was considered for pattern vectors that did not participate in the training. The expectation was taken *both* with respect to $2N$ random training vectors and with respect to a random character of generating the $(p+1)$-variate weight vector determined by the prior density $f_{prior}(\mathbf{w}, w_o) = N(0, I)$. Numerical results calculated by integrating a certain double integral are obtainable roughly from the following approximate formula for the ZEE classifier with random priors:

$$MEP_N^{(ZEE)} \approx \Phi\{-\frac{\delta}{2}\frac{1}{\sqrt{1+(1.6+0.18\delta)(p/N)^{1.8-\delta/5}}}\}. \tag{14}$$

In addition, a model was analyzed where in the random search training procedure one selects hyper-planes such that classify the learning sets LS^1 and LS^2 without error, and the margin (the Euclidean distance) between the hyper-plane $w'x + w = 0$ and the learning set vector x^* closest to it exceeds Δ, the value of the bound for the margin. It is the margin classifier. It has been shown theoretically that an increase in Δ diminishes the MEP_N, the average in *different* parts of learning sets. The simulation with Gaussian patterns have shown, however, that for each particular training set an excessive growth of the margin increases the generalization error. Thus, we can conclude that the maximal margin classifier is the most complex one from a series of statistical classifiers that can be obtained in non-linear SLP training.

Equation (14) and the numerical results indicate that a non-parametric approach for designing a linear classifier allows us to obtain reliable rules even in the cases where the number of features is notable larger than the number of training vectors. Rejection of the assumptions that the classes are Gaussian leads to refusal from the estimation of class global parameters Σ and μ_1, μ_2, the covariance matrix, and the means. The estimation of these parameters in a high-dimensional case is not favourable for classifier design. In the large learning-set situations, however, the parametric Fisher classifier is less sensitive to the learning-set size. Thus, one can affirm that, in the large learning-set situations, the minimum empirical error classifier is more complex than the parametric Fisher classifier, and that *in SLP training we move from the simplest Euclidean distance classifier to the more complex regularized discriminant analysis, afterwards to the Fisher DF, and to the minimum empirical error and the maximal margin classifiers.*

5. OPTIMAL COMPLEXITY OF THE NON-LINEAR SLP CLASSIFIER

A transition from the simplest classifier to more complex classifiers is one of the reasons that cause the overtraining effect. The earlier analysis of the generalization error and small sample properties of statistical classifiers does not answer the question which type of classifier to choose, what learning parameters have to be, how many iterations to use in each particular situation. Naturally the answer depends on the data, its size and configuration. We will explain the situation by presenting three simple examples with artificial bi-variate data sets.

Example A. If we have spherical Gaussian classes, then we have to use the EDC which is the optimal sample-based classifier for this type of data. It means that *the SLP should be trained only by one iteration.* An increase in the number of iterations will lead to different classifiers that would yield higher generalization error This theoretical conclusion is vividly illustrated in Fig. 6 a,b. We have used here 2-variate spherical Gaussian data with $\delta=6.6$ (the Bayes error $P_B =0.000483$). The learning set size $N_1= N_2=200$ is sufficiently large to train the EDC. Therefore, after the first iteration we have obtained the generalization error P_{gen}=0.000488, only slightly higher than P_B. Four hundred training iterations with the learning step η=2 increased the generalization error to 0.00056 (Graph. 1 in Fig. 6a).

In this experiment we have no misclassifications in the learning set. So the minimum of the cost function was obtained with very large weights. With an increase in the magnitude of weights all outputs $o(\mathbf{w}'\mathbf{x}_j^{(i)})$ become close to +1 or to -1, and the gradient of the cost function then is close to zero. Therefore in the back propagation (BP) training with fixed learning step η the training process became very slow.   In order to force the back propagation algorithm change the weights uniformly and increase the margin, one has to increase the magnitude of the learning step.

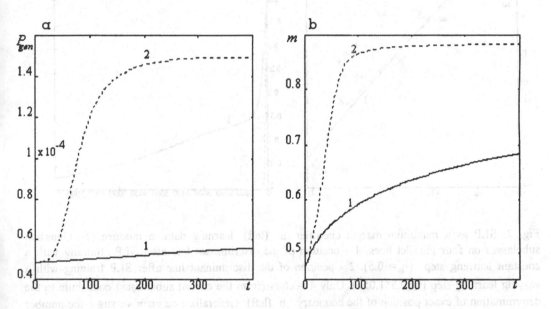

Fig. 6. a (left): Generalization error as a function of the number of iterations t. b (right): The margin as a function of t.   1 - η = 2, 2 - η = 2*1.1^t.   Spherical Gaussian data; p=2, δ=6.6, N= 200.

The analysis shows that to ensure a linear increase in the magnitude of weights, the learning step η should progressively increase with the increase in iterations number t:

$$η = η_{start} * α^t.$$ (15)

where α is a positive constant, a little larger than 1.

Therefore, in the next experiment, we have changed η according to (15) with $η_{start}$ =2 and α=1.1. The usage of this learning strategy allowed us to obtain the maximum margin classifier fast, but it caused an enormous overtraining effect, the generalization error has increased more than 3 times! (see graph 2 in Fig. 6 a and b).

   Example B. The previous example has demonstrated one extreme case where we have to train the SLP only by one iteration. Now we present *the opposite example*, where it is necessary to train the SLP maximally. Each class here consists of two subclasses. It is a mixture of two Gaussian densities. Each subclass is distributed on a straight line in the bi-variate space. All the four lines are parallel (see Fig.7a).

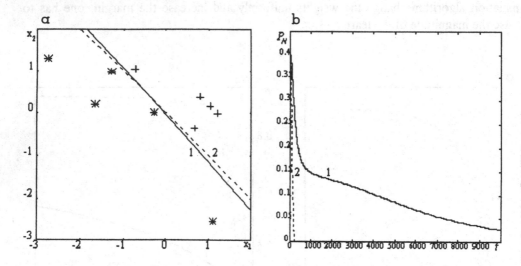

Fig. 7. SLP as a maximum margin classifier. a (left): learning data: a mixture of Gaussian subclasses on four parallel lines; 1 - position of the discriminant line after SLP training with a constant learning step $(\eta = 0.5)$. 2 - position of the discriminant line after SLP training with a varying learning step $(\eta=0.5*1.05^{t})$. Only 4 vectors (from the closest subclasses) contribute to the determination of exact position of the boundary. b (left): Generalization error versus t, the number of iterations: 1 - $\eta = 0.5$, 2 - $\eta=0.5*1.05^{t}$.

   After SLP training with the progressively increasing learning step, the decision boundary was situated approximately halfway between the vectors of two closest subclasses of the opposite classes (line 2 in Fig. 7a). It is the maximum margin classifier. The remaining vectors from the other two subclasses do not affect the position of the decision boundary. Use of conventional training with the constant value of the learning step was extremely slow and did not favour to obtain the maximal margin and good generalization: after 10 000 iterations the margin was 0.18 and the generalization error 0.026. The use of the progressively increasing learning step enabled us to obtain the maximal margin ($m=0.21$) and zero empirical error just after 250 iterations (see Fig. 7b). For such a configuration of data, the maximal margin classifier is the best type of the classification rule and in order to obtain it one needs to train the SLP as long as the minimum of the cost function is reached.

Example C. In the previous section, we have demonstrated that the SLP weights depend on the seperability of pattern classes. When the empirical error is high, we have small weights and will never obtain a minimum empirical error classifier. In order to force the classifier to minimize the empirical frequency of misclassifications, we may add to cost function (3) a *"negative weight decay"* term, so called *"anti-regularization"* - $\lambda_1 w'w$ or $+\lambda_2(w'w/h^2-1)$ [6]. Here, the anti-regularization term will force to increase the magnitude of weights and, consequently, to increase an actual slope of the activation function.

This technique is illustrated in Fig. 8ab. In Fig. 8a, we present the distribution of two bi-variate Gaussian classes $N(\mu_i, \Sigma)$ - two small ellipses - contaminated with 10% additional Gaussian noise $N(0, N)$; $N=500$. The noise patterns are denoted by "stars" and "pluses" and the signal classes - by two ellipses. Both signal classes have different means $\mu_1 = -\mu_2 = \mu$ and share a common covariance matrix Σ:

$$\mu = \begin{bmatrix} 0.030 \\ 0.015 \end{bmatrix}, \quad \Sigma = \begin{bmatrix} 0.040 & 0.0196 \\ 0.0196 & 0.01 \end{bmatrix}, \quad N = \begin{bmatrix} 1.0 & -0.7 \\ -0.7 & 1.0 \end{bmatrix}.$$

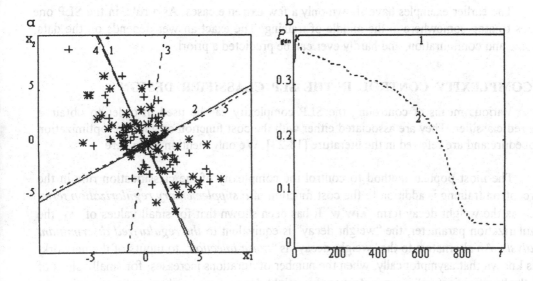

Fig. 8. a: Distribution of two Gaussian pattern classes contaminated with additional noise, and positions of the discriminant boundaries. b: Generalization error versus t, the number of iterations. 1 - SLP, conventional training, 2 - SLP, with anti-regularization. 3 - EDC, 4 - Fisher linear DF, 5- optimal linear DF.

We'd like to draw reader's attention to the fact that in this data, the components of "signal" and "noise" vectors have the opposite correlations. The variance of noise is much larger than that of a signal. Therefore the direction of the decision boundary of the Euclidean distance classifier (Graph 3) and that of the Fisher classifier (Graph 4) substantially differ from that of the boundary of the optimal linear classifier (Graph 5). In our traditional training ($t_j^{(1)} = 1$, $t_j^{(2)} = -1$, $\overline{x}^{(1)} + \overline{x}^{(2)} = 0$, $N_2 = N_1$, $w_{0(0)} = 0$, $\mathbf{w}_{(0)} = 0$, $\eta = 10$), the boundary of the SLP moves from (3) towards (4) and settles at (1). Graph 1 (after 1000 iterations) is close to the decision boundary of the Fisher DF. After the first iteration with $\eta = 10$, we have the generalization error 0.34, which, when approaching the Fisher classifier, increases up to 0.39 (Fig. 8b, Graph 1).

Addition of a supplementary anti-regularization term in the cost function of the sum of squares $+10000(\mathbf{w'w}/25^2 - 1)^2$ does not change the learning curve at the very beginning, but later on, when the weights increase substantially, the decision boundary begins approaching the optimal boundary. Graph 2 in Fig. 7a shows the position of the boundary after 1000 iterations. As a result, we obtain small generalization error $P_{gen} = 0.043$, the same as the value obtained by using the optimal linear classifier designed to classify only the Gaussian "signal" patterns.

The earlier examples have shown only a few extreme cases. As a rule, in the SLP one needs to stop somewhere in the middle of training. The exact answer depends on the data, its size and configuration, and hardly ever can be predicted a priori.

6. COMPLEXITY CONTROL IN THE SLP CLASSIFIER DESIGN

Various means of controling the SLP complexity can be used in order to obtain a desired classifier. They are associated either with the cost function or with the optimization procedure and are referred in the literature [18-23]. We only mention them here.

The most popular method to control the complexity of the classification rule in the perceptron training is addition to the cost function of *a supplementary regularization term*, such as the weight decay term $\lambda_1 \mathbf{w'w}$. It has been shown that for small values of λ_1, the regularization parameter, the "weight decay" is equivalent to *the regularized discriminant analysis*. An alternative to the "weight decay" is "*noise injection*" to inputs of the network. It is known that asymptotically, when the number of iterations increases, for small values of λ_1, the "noise injection" is equivalent to the weight decay. Some authors suggest adding the noise to the weights or even to the outputs of the network. Earlier in the third section, we have shown that, at the very beginning of the training process, the weight vector $\mathbf{w}_{(t)}$ of the SLP is equivalent to that of the regularized discriminant analysis with the regularization parameter $\lambda_1 = \dfrac{2}{(t-1)\eta} \dfrac{N}{N-1}$. Therefore *the number of iterations* acts as the regularized too. In order to obtain a wider range of classifiers, instead of adding the regularization term we have proposed a supplementary "negative weight decay" term called *"anti-regularization"*. This technique is effective when we have a great percentage of outlyers - atypical observations - and the empirical classification error is high.

The learning step η can also serve as a complexity control factor. E.g., a very high value of the learning step η can lead to a situation, where, after a first learning iteration, the activation function will be saturated and further training will not be possible. Then we obtain only one classifier - EDC. Comparatively small values of η slow down the training process, and in the case of zero empirical error, to obtain the maximum margin classifier we suggested using *a progressively increasing learning step*. When training with a small learning step, one always obtains a regularized DA and approach the Fisher classifier. However, if η is too high, then, in the zero empirical error case, one will skip the Fisher DF with a pseudo-inversion and will go straight in the direction of the maximum margin classifier.

After a certain number of iterations, the training of SLP moves the decision boundary from that of the Fisher classifier to the generalized DA and farther towards the minimum empirical error classifier. This process can be controlled by *changing the target values*. If we use limit values of the targets (e.g., +1 and -1 for activation function (2)), then, in principle, we can obtain large weights and a classifier similar to the minimum empirical error classifier. If we have non-limit values of the targets (e.g. +0.5 and -0.5 for activation function (2)), then the weights will never be large and we will not move towards the minimum empirical error classifier. Thus, the target values can also act as regularizers.

One more factor traditionally used to control the complexity is selection of a proper *architecture of the network*. There are two typical strategies of doing this. One of them is *incremental learning*. Here, in the training process, the network size is sequentially increased by adding new neurones or new features (inputs). The another strategy is *pruning* the network when the neurones or features are sequentially rejected.

All the factors enumerated act *simultaneously*, and frequently (not always) the influence of one factor can be compensated by another one. There are several directly uncontrolled factors. These are *false local minima* and *high-dimensional extremely flat areas of the cost function*, where the training process actually stops. Adding noise to inputs of the perceptron or its weights as well as a constant or a temporal increase in the learning step can help to move the perceptron weight vector from these unlucky areas.

In each training experiment, one cannot obtain all the seven classifiers enumerated. If *the conditions of obtaining EDC* are fulfilled (batch-mode training, $\overline{x}^{(1)} + \overline{x}^{(2)} = 0$, $w_{0(0)} = 0$, $\mathbf{w}_{(0)} = 0$, $t_2 = -t_1 N_1 / N_2$,), then, after the first BP step, we always obtain the Euclidean distance classifier. Otherwise, we will not. Thus, the weight initialization as well as data normalization affect the type of classifier obtained too. Further training results, however, depend on the data, the cost function, and training conditions.

7. CONCLUSIONS

There are two main reasons that cause the overtaining effect in the perceptron training.

1. Due to the differences in the learning set and the test set, we have different surfaces of learning-set and test-set cost functions in the multivariate perceptron weights space. Moreover, in the classification problem, we minimize the smooth and differentiable cost function whereas actually we are interested in the classification error.

2. In the way between the starting point and the minimum of the cost function, the non-linear activation function leads to a situation where, while traning the linear adaptive classifier, with an increase in the magnitude of weights, the cost function is gradually changing.

The second reason can lead to the seven statistical classification rules of different complexity. After a first learning iteration we can obtain the simplest Euclidean distance classifier, and at the very end of training we can obtain the most complex maximal margin classifier. Between these two classifiers one can obtain the regularized discriminant analysis, the Fisher linear discriminant function, the Fisher linear discriminant function with the pseudo-inversion, the generalized Fisher discriminant function, and the minimum empirical error classifier.

Therefore, the overtraining preventing problem can be analyzed as the problem of selection of the proper type of statistical classifier. *Which classifier is the best one for a given situation depends on the number of features, data size and its configuration.*

For spherical Gaussian classes, the best strategy is to stop training after the first iteration, when the EDC is obtained. For linearly separable classes with sharp linear boundaries of distributions, we have to train the perceptron as long as the minimum of the cost function is obtained. For intermediate and more realistic situations, usually we have to stop earlier.

In order to determine an optimal stopping moment we have to use additional information never used before in training the perceptron. In high-dimensional cases, assumptions on the probabilistic structure of data and consequent theoretical considerations can help estimate the average value of the optimal stopping moment. Use of an additional validation set is a universal method. It is, in fact, a special way to extract the useful information contained in the validation set.

Various control techniques of different complexity can be used to obtain the proper type of classifier. In some cases, however, the conventional training methods do not lead to the best type of classifier. In order to obtain the minimal empirical error or the maximal margin classifiers easier and faster one needs to apply special techniques such as a progressively increasing learning step in BP training an/or a supplementary anti-regularization term added to the conventional cost function of the sum of squares. A proper choice of parameters of these two procedures is the object of further investigations.

ACKNOWLEDGEMENTS

The author thanks Valdas Diciunas, Robert P.W. Duin and Francoise Fogelman-Soulie for stimulating and fruitful discussions and aid.

REFERENCES

1. Rumelhart D.E., G.E.Hinton and R.J.Williams: Learning Internal Representations by Error Propagation, in: Parallel distributed processing: Explorations in the microstructure of cognition, vol. I, Bradford Books, Cambridge, MA, 1986, 318-362.

2. Cramer, G.: Mathematical methods of statistics, Princeton University Press, Princeton, New York 1946.

3. Raudys S.: Linear classifiers in perceptron design. Proceedings 13th ICPR, Track D, August 25-29, 1996, Viena, IEEE Publication.

4. Sebestyen, G.S.: Decision-making process in pattern recognition, Mcmillian, NY 1962.

5. McLachlan G.: Discriminant Analysis and Statistical Pattern recognition, Willey, NY 1992.

6. Raudys S.: A negative weight decay or antiregularization, Proc. ICANN'95, Oct. 1995, Paris, Vol. 2, 449-454.

7. Randles, R.H., J.D. Brofitt, I.S. Ramberg and R.V. Hogg: Generalised linear and quadratic discriminant functions using robust estimates. J. of American Statistical Association, 73, (1978), 564-568.

8. Cortes, C. and V.Vapnik: Support-vector networks, Machine Learning, 20, (1995), 273-297.

9. Raudys, S.: On determining training sample size of a linear classifier, in: Computing Systems, 28 (Ed. N. Zagoruiko), Institute of Mathematics, Academy of Sciences USSR, Nauka, Novosibirsk, (1967), 79-87 (in Russian).

10. Deev, A.D.: Representation of statistics of discriminant analysis and asymptotic expansions in dimensionalities comparable with sample size, Reports of Academy of Sciences of the USSR, **195**, No. 4, (1970), 756-762, (in Russian).

11. Raudys, S.: On the amount of a priori information in designing the classification algorithm, Proc. Acad. of Sciences of the USSR, Technical. Cybernetics, N4, Nauka, Moscow, (1972), 168-174 (in Russian).

12. Raudys, S.: On the problems of sample size in pattern recognition, Proc. 2nd All-Union Conf. Statistical Methods in Control Theory (Ed. V.S. Pugatchev), Nauka, Moscow, (1970), 64-67 (in Russian).

13. Kanal L. and B.Chandrasekaran: On dimensionality and sample size in statistical pattern recognition, Pattern Recognition, **3**, (1971), 238-255.

14. Jain, A. & Chandrasekaran, B.: Dimensionality and sample size considerations in pattern recognition practice, Handbook of Statistics, **2**, North Holland, 1982, 835-855.

15. Raudys, S. and A.K. Jain: Small sample size effects in statistical pattern recognition: Recommendations for practitioners, IEEE Trans. on Pattern Analysis and Machine Intelligence, **PAMI-13**, (1991), 252-264.

16. Raudys, S., M.Skurikhina, T.Cibas, P.Gallinari: Ridge estimates of the covariance matrix and regularization of an artificial neural network classifier, Pattern Recognition and Image Processing, Int. J. of Russian Academy of Sciences, Moscow, No 4, 1995.

17. Raudys, S. and V. Diciunas V.: Expected error of minimum empirical error and maximal margin classifiers, Proceedings 13th ICPR, Track **B**, August 1996, Wien. IEEE Publ.

18. Mao, J. and A.Jain: Regularization techniques in artificial neural networks, Proc. World Congress on Neural Networks, July 1993, Portland.

19. Reed R.: Pruning Algorithms - A Survey, IEEE Trans on Neural Networks, **4**, (1993), 740-747.

20. Bishop C.M.: Regularization and complexity control in feed-forward networks. Proceedings ICANN'95, **1**, Oct. 1995, Paris, (1995), 141-148.

21. Canu S.: Apprentissage et approximation: les techniques de regularisation. Cours de DEA, Chapitre 1, Univ. Technologie de Compiegne, 1995.

22. Reed R., R.J. Marks II, S.Oh: Similarities of error regularization, sigmoid gain scaling, target smoothing, and training with jitter, IEEE Transaction on Neural Networks, 6, (1995), 529-538.

23. Raudys, S. and F.Fogelman-Soulie: Means of controling the complexity in ANN classifier design. An overview. Submitted to print, 1996.

NEURAL NETWORKS FOR RAPID LEARNING IN COMPUTER VISION AND ROBOTICS

H. Ritter

Bielefeld University, Bielefeld, Germany

1 Introduction

One of the major thrusts for the success of neural networks in many areas was the development of learning algorithms that allow to capture implicit knowledge that is only available in the form of examples such that it can be generalized and applied to new situations, usually "inputs" to a suitably trained net.

In the past, the majority of efforts was concentrated on multilayer perceptrons, and a number of algorithms for supervised training of such systems has been developed. A distinguishing feature of these systems is their highly distributed representation which is favorable for good generalization and for robustness against noise. However, it also gives rise to problems, most notably the difficulty to guarantee good convergence of the training procedure and the interference between new and previously trained data which make the implementation of incremental learning schemes difficult.

Issues like these are more easily handled in more "localist" approaches, such as radial basis function networks ([MD88, GJP95]) or, more recently, support vector nets ([CV95]). These approaches have close relationships to kernel regression methods and offer a great deal of flexibility for the tuning of their approximation capabilities by making suitable choices for a regularization function or the form of a scalar product in feature space.

In this contribution, we want to focus on a class of neural learning algorithms which can be viewed as located between the two extremes of purely localist approaches, such as RBF-networks, and fully distributed approximation nets, such as multilayer perceptrons. These algorithms have been derived from the self-organizing map, which originally was motivated by the desire to model the structuring of neural layers in the brain.

The self-organizing map [Koh90] provides a non-linear method by which a layer of adaptive units is gradually transformed into an array of local feature detectors that capture statistically important features of their input signals. Therefore, the original version of the self-organizing map can be viewed as belonging into the class of localist models, although there is a notion of topological neighborhood among the local units that introduces a non-local element by coupling the adaptation steps of neighboring units. However, the local nature of the feature detectors has so far limited the approach to signal distributions with a low-dimensional variance. In this paper, we discuss two extensions of the basic self-organizing map that allow to overcome this limitation to a considerable extent and that still can be used when the data distribution fills a higher dimensional manifold.

2   Local Linear Maps

The first extension, the *Local Linear Map*-networks (LLMs), replace the units of the standard self-organizing map by *locally valid linear mappings* [RS87, RMS92, MR92]. This combines the many virtues of linear models with the ability to handle nonlinearity by developing a suitable tesselation of the underlying feature space such that in each tesselation cell a linear approach becomes feasible. A similar suggestion, in the context of time-series prediction, has also been made by [SU90]. There have also been attempts to introduce some degree of locality into multilayer perceptrons by replacing one large net by a collection of smaller "expert nets" whose outputs are suitably combined by a superordinate "gating net" [JJNH91]. More recent developments of this approach have now replaced the expert nets by linear mappings [JJ94], thereby coming very close to the local linear map approach.

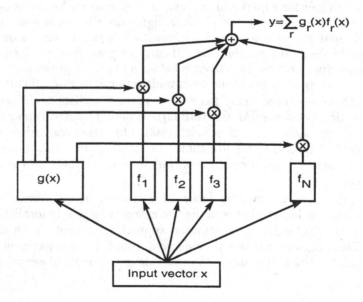

Figure 1: Local Linear Map Network

A LLM-network [RMS92, MR92] (schematically shown in Fig.1) consists of a single internal layer of N units, labelled by an index r. Each unit receives the same input vector $\mathbf{x}$ of dimension L and implements a locally valid linear mapping, specified by two vectors $\mathbf{w}_r^{(in)} \in \mathbb{R}^L$ ("input weight vector"), $\mathbf{w}_r^{(out)} \in \mathbb{R}^M$ ("output weight vector") and a $M{\times}L$-matrix $\mathbf{A}_r$ (M is the dimensionality of the output space). The output $\mathbf{y}_r$ of a unit is given by

$$\mathbf{y}_r = \mathbf{w}_r^{(out)} + \mathbf{A}_r(\mathbf{x} - \mathbf{w}_r^{(in)}).$$

This represents a set of linear mappings with Jacobians $\mathbf{A}_r$, passing through the points $(\mathbf{x}, \mathbf{y}) = (\mathbf{w}_r^{(in)}, \mathbf{w}_r^{(out)}), r = 1 \ldots N$. Which of these mapping is used to obtain the output $\mathbf{y}^{(net)}$ of the network is determined by the distances $d_r = \|\mathbf{x} - \mathbf{w}_r^{(in)}\|$ between the input $\mathbf{x}$ and the input weight vectors $\mathbf{w}_r^{(in)}, r = 1 \ldots N$. In the simplest case, the mapping associated with the unit s for which this distance is minimal, i.e. for which $d_s = \min_r d_r$, is used ("winner-take-all"-network). However, this introduces discontinuities at the borders of the Voronoi tesselation cells given by the $\mathbf{w}^{(in)}$ vectors. Such discontinuities can be avoided by introducing a "soft-max"-function

$$a_s = \frac{\exp(-\beta d_s)}{\sum_r \exp(-\beta d_r)} \tag{1}$$

with an "inverse temperature" $\beta > 0$ and blending the contributions of the individual nodes according to

$$\mathbf{y}(\mathbf{x}) = \sum_r a_r \cdot [\mathbf{w}_r^{(out)} + \mathbf{A}_r(\mathbf{x} - \mathbf{w}_r^{(in)})].$$

A simple learning rule can be used to train the prototype vectors $\mathbf{w}_r^{(in)}$, $\mathbf{w}_r^{(out)}$ and the Jacobians $\mathbf{A}_r$ from (labelled) data examples:

$$\Delta \mathbf{w}_r^{(in)} = \epsilon_1(\mathbf{x} - \mathbf{w}_r^{(in)}) \tag{2}$$

$$\Delta \mathbf{w}_r^{(out)} = \epsilon_2(\mathbf{y} - \mathbf{y}^{(net)}(\mathbf{x})) + \mathbf{A}\Delta \mathbf{w}_r^{(in)} \tag{3}$$

$$\Delta \mathbf{A}_r = \epsilon_3(\mathbf{y} - \mathbf{y}^{(net)}(\mathbf{x}))\frac{\mathbf{x}^T}{\|\mathbf{x}\|^2} \tag{4}$$

Here, the learning rule for output values $\mathbf{w}_r^{(out)}$ differs from the equation for $\mathbf{w}_r^{(in)}$ by the extra term $\mathbf{A}\Delta \mathbf{w}_r^{(in)}$ to compensate for the shift $\Delta \mathbf{w}_r^{(in)}$ of the input prototype vectors. The adaptation rule for the Jacobians is the perceptron learning rule, but restricted to those input-output pairs $(\mathbf{x}, \mathbf{y})$ such that $\mathbf{x}$ is in the Voronoi cell of center $\mathbf{w}_r^{(in)}$ (in practice, one should decrease the learning rates $\epsilon_{1\ldots3}$ slowly towards zero, with ϵ_3 decreasing more slowly than the other two in order to leave any fine-tuning to the matrices $\mathbf{A}_r$).

Due to the local linear mappings, the number of required centers $\mathbf{w}_r^{(in)}$ often can be rather low (say, some tens). In these cases, the approach can be considered as a natural extension of a single linear model, with the nonlinearity nicely "encapsulated" in equation 1 that determines the relative weighting of the local linear maps used.

A natural way to extend this scheme might be to increase the approximation order further and to use locally quadratic functions. This, however, leads to a steep increase in the number of to-be-determined parameters (a quadratic form in d-dimensional space has $d(d+1)/2$ parameters instead of only d for the linear case). A somewhat more moderate step is then to abandon the use of many local higher-order models and instead represent the *entire mapping* in a suitably parametrized form. This approach leads to the parametrized self-organizing maps and is described in Sec.4; before, we shall briefly discuss an example of an application of LLM-networks to a computer vision task.

3   Computer Vision Applications

We have applied LLM-networks to a variety of pattern recognition tasks, particularly in the domain of computer vision (see, e.g, [MR92, HKRS96]). Here, the input data are images that must first be reduced to a sufficiently low-dimensional feature vector in order to allow training of a network. In many cases, a first preprocessing step is the segmentation of the image to filter out background or unwanted image contents. In practical applications, this segmentation step can be very crucial; since it is a classification task, it can in principle again be approached with LLM-networks (ref. [LR97] presents a comparison between LLM networks and other classification methods for color based pixel classification). In pixel classification, however, the resolution is often limited and then a significant gain in speed can be achieved by storing the LLM-classifier in discretized form in a table of input-output pairs. Here, we will not consider the segmentation step (for the use of LLM-networks for this task, the reader is referred to [LR97]) and only present an example for the use of LLM-networks to identify particular object points in images that show a single object that is well-separated from the background.

A particularly interesting "object" is the human hand. From the perspective of computer vision, recognizing hands is a challenging task, since hands are non-rigid objects of a highly variable and rather complex shape. For obvious reasons, recognizing hands is also of considerable practical interest, e.g., to create novel interactive input devices for computers.

A full recognition of the rich variety of different postures that a human hand is able to attain is currently beyond the abilities of any existing computer vision algorithm and will almost certainly require a subdivision of this very complex recognition task into a number of more manageable subtasks. One such subtask is the identification of the locations of prominent spatial features of the hand, such as the finger tips.

We can approach this task by labelling a suitable set of images of human hands with the correct finger tip positions and then use these images as training examples for LLM-networks to identify the labeled finger tip positions from the image intensity pattern.

However, to make this approach feasable requires to reduce the formidably high-dimensional intensity vector of the raw pixel intensity values to a feature vector of a more manageable size. Different schemes for this dimension reduction step can be used. A very suitable procedure (which is also known to have parallels with the processing of retinal signaly by cortical cells in the primary visual cortex, cf. [Dau80]), consists

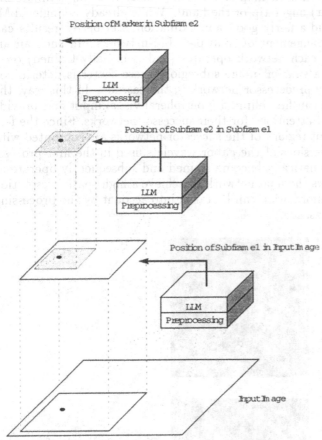

Position of Marker in Subfram e2

LLM
Preprocessing

Position of Subfram e2 in Subfram e1

LLM
Preprocessing

Position of Subfram e1 in Input Im age

LLM
Preprocessing

Input Im age

Figure 2: Cascaded arrangement of LLM-networks to identify finger tip positions in image. Each network determines a smaller image region that then serves as input for the next network. The final network outputs the position estimate.

in first forming an edge-enhanced image by convolving the raw intensity pattern with a spatial high-pass filter, such as a laplacian, and subsequently forming a set of cross-correlation coefficients, each one resulting from cross-correlating the image intensity with a different spatial filter kernel, such as, e.g. a gabor wavelet (the high-pass filtering can be omitted, if the chosen gabor wavelets include a sufficient number of filters with high wave number; however, since we here use only a very sparse subset of all possible gabor wavelets, the high-pass filtering is useful[1]).

These preprocessing steps yield for each training image a feature vector with one image feature per gabor wavelet. For the present task, we used a set of 36 gabor wavelets, centered at the points of a quadratic 3×3 lattice (using for each point the same spatial resolution and 4 different orientations, separated by angles of 90^0), covering a central region of the image area.

---

[1] since the high-pass filter and the cross correlation are both linear operations, they can be combined into a single filtering step with correspondingly modified filter kernels.

The task of the LLM-network is to map this feature vector into (an estimate of) the 2d-location of a (particular) finger tip of the hand. While already a single LLM-network can be trained to yield a fairly good approximation, still better results can be obtained by a cascaded arrangement of multiple LLM-networks. In such an arrangement, depicted in Fig.2, each network operates (and is initially trained) on a feature vector extracted from a smaller image subregion whose center is determined by the position estimate of the predecessor network in the cascade. In this way, the initial networks in the cascade can first eliminate peripheral visual input and provide increasingly focused "regions of attention" for their successor networks. Since the feature vectors in the smaller input regions of the successor networks are computed with correspondingly down-scaled versions of the gabor wavelets used in the first processing stage, these networks automatically become trained and subsequently operate at a finer spatial scale. This allows the final network to achieve a high spatial resolution while still more global visual information can be taken into account by the processing of the initial networks in the cascade.

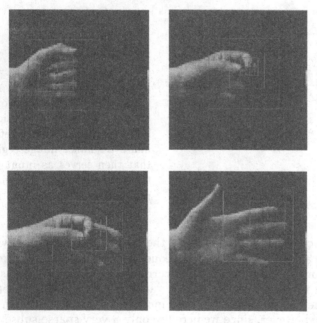

Figure 3: Result of applying the cascaded networks of Fig. 2 to test images of different hand shapes. In this case, the task was to identify the position of the index finger in the image.

Fig.3 shows some results of such an approach for locating the index finger of a human hand. In this case, we used the (rather low) number of 50 training images. The depicted recognition examples are some typical cases from an independent test set of 50 further images (showing the same hand and using the same illumination conditions; at present, we did not yet provide any measures for normalizing illumination or the size of the hand). We are currently extending this approach for recognizing multiple finger tips simultaneously, including a more robust preprocessing and segmentation stage and considering the evaluation of context information that becomes available when several finger tip positions are known or when continuous motion sequences are considerd.

4 Parametrized Self-organizing Maps

The degree of locality of the LLM-approach is essentially given by the size of its tesselation cells. Compared with more distributed representations, such as the well-known multilayer perceptron, the locality allows to confine local changes to the vicinity of a data point, such that incremental re-adaptation with new data becomes possible without having to re-train on the entire data set, as is usually necessary for multilayer-perceptrons.

If a more distributed representation is desired, this can be achieved by increasing the order of the local approximation scheme while at the same time making the cells larger in order to avoid overfitting. However, higher order approximation schemes for general tesselation cells are not very convenient to handle. A workable possibility is to insist on a simple, say rectangular, structure of the tesselation cells in some "latent" parameter or "map" space S and to use now two smooth functions $w^{(in)}(s)$ and $w^{(out)}(s)$ to map S into the input space X^{in} and output space X^{out}, resp. This makes S into a manifold that is embedded in the cartesian product space $X^{in} \times X^{out}$ and that can be viewed as the graph of a (possibly multi-valued) function $X^{in} \mapsto X^{out}$.

The rectangular tesselation on S is used to represent the mappings $w^{(in)}$ and $w^{(out)}$ by some convenient interpolation scheme (for instance, using some set of polynomials that are orthogonal on a tesselation grid $\tilde{A}$ chosen on S) in terms of the discrete set of values $w^{(in)}$ $w^{(out)}$ that occur at the discretization points s in S.

Such scheme can be viewed as a self-organizing map with a continuous "map manifold" S and may thus be called a *parametrized self-organizing map*. In contrast to the standard SOM (where S is given in discretized form by a set of "reference" or "weight" vectors "attached" to a d-dimensional lattice [Koh90, RMS92]) in a PSOM S is given as a parametrized hypersurface $(w^{(in)}(s), w^{(out)}(s))$.

Considerable additional flexibility can be gained by not insisting on a fixed determination of which subspace is input and which is output space but instead formulating everything in terms of a concatenated vector $w(s) = (w^{(in)}(s), w^{(out)}(s))$. By introducing a set of scalar valued basis functions $H(s, a)$ that obey the conditions

$$H(a, a') = \delta_{a,a'} \quad \forall a, a' \in \tilde{A} \tag{5}$$

$$\sum_{a \in \tilde{A}} H(a, s) = 1 \quad \forall s \in S \tag{6}$$

one may write the mapping from manifold parameters s into $X = X^{in} \times X^{out}$ as

$$w(s) = \sum_{a \in \tilde{A}} H(s, a) w_a. \tag{7}$$

where s is a d-dimensional vector of coordinates on the hypersurface. The vectorial coefficients w_a are the adaptable parameters and play the role of support points through which the hypersurface shall pass. Like in a SOM, each w_a is "attached" to a node a in a d-dimensional discrete (and usually hypercubic) lattice $\tilde{A}$, over which the summation runs. The second requirement, Eq. (6), guarantees that the resulting manifold $w(s)$ does not change its shape under a translation $w_a \rightarrow w_a + \vec{t}$ of all reference vectors.

When S is specified, the PSOM is used in an analogous fashion like the SOM: given an input vector $\mathbf{x}$, (i) first find the bestmatch position $\mathbf{s}^*$ on the manifold S by minimizing a distance $d(\mathbf{x}, \mathbf{w}(\mathbf{s}))$, and then (ii) use the surface point $\mathbf{w}(\mathbf{s}^*)$ as the output of the PSOM in response to the input $\mathbf{x}$.

Usually, $\mathbf{x}$ and $\mathbf{w}(\mathbf{s}^*)$ will be from the same vector space $\mathbf{X}$, and it is understood that only a subset of the components of $\mathbf{x}$ must be specified and is regarded as input. This is implemented by specifying a subset I of the index values $1 \ldots n$ $(n = \dim(\mathbf{X}))$ and defining the distance d by

$$d(\mathbf{x}, \mathbf{w}(\mathbf{s})) = \sum_{i \in I} (x_i - w_i(\mathbf{s}))^2, \tag{8}$$

i.e., only vector components selected by the index set I contribute and can, therefore, influence the choice of the resulting bestmatch vector $\mathbf{y} = \mathbf{w}(\mathbf{s}^*)$. We identify the subspace that is spanned by these components with the input space $\mathbf{X}^{\text{in}}$ introduced above. Those components of $\mathbf{y} = \mathbf{w}(\mathbf{s}^*)$ with indices not from I will then span the orthogonal complement $\mathbf{X}^{\text{out}}$ and can be considered as outputs that provide values for the missing or ignored components of the input vector $\mathbf{x}$. Thus, the hypersurface given by the vector function $\mathbf{w}(\mathbf{s})$ acts as the *graph* of a mapping $\phi : \mathbf{X}^{\text{in}} \mapsto \mathbf{X}^{\text{out}}$.

To make $\mathbf{w}(\mathbf{s})$ represent a hypersurface that passes through a number of desired data pairs $(\vec{\xi}, \vec{\eta})^\nu =: \mathbf{w}^\nu$ (note that each data pair $(\vec{\xi}, \vec{\eta})^\nu$ is represented by a *single vector* $\mathbf{w}^\nu \in \mathbf{X}$) requires a suitable choice of basis functions $H(\mathbf{a}, \cdot)$ and vectorial coefficients $\mathbf{w}_\mathbf{a}$. A particularly simple choice results if we assume that the data points $\mathbf{w}^\nu \in \mathbf{X}$ can be assumed to result from sampling ϕ on a (possibly distorted) d-dimensional sampling grid $\tilde{A}$ (cf. eq. (7)). Then we can always construct a set of basis functions with the properties

$$H(\mathbf{a}, \mathbf{a}') = \delta_{\mathbf{a}, \mathbf{a}'} \quad \forall \mathbf{a}, \mathbf{a}' \in \tilde{A} \tag{9}$$

on the subset of lattice locations $\tilde{A}$ (see, eg. [Rit93, WR96b]). Substituting eq. (9) into eq. (7) yields $\mathbf{w}(\mathbf{s}) = \mathbf{w}_\mathbf{s}$ for all $\mathbf{s} \in \tilde{A}$. Therefore, if $\mathbf{w}^\nu$ is the data point that results from sampling ϕ at lattice point $\mathbf{a}$, setting $\mathbf{w}_\mathbf{a} = \mathbf{w}^\nu$ immediately ensures that the hypersurface passes through this data point. Therefore, we can immediately construct a PSOM that is *exactly correct* at the specified data points, simply by choosing the (possibly relabeled) given data points $\mathbf{w}^\nu$ for the vectorial coefficients $\mathbf{w}_\mathbf{a}$ in (7). The generalization error on new, additional points then depends on the extent to which the interpolation by the parametric surface matches the true underlying function.

The price to pay for this very direct and simple construction of the PSOM is an increased cost when using it. For step (i), the best match parameter vector $\mathbf{s}^*$ is defined by

$$\mathbf{s}^* = arg \min_{\mathbf{s}} d(\mathbf{x}, \mathbf{w}(\mathbf{s})) \tag{10}$$

Inspection of eq. (10) shows that the discrete bestmatch search for the standard SOM case has now been replaced by a nonlinear minimization problem, which must be solved iteratively. The simplest method is iterative gradient descent, e.g., from a starting value $\mathbf{s}(t = 0)$ iteratively determine a sequence of values $\mathbf{s}(t)$, $t = 1, 2 \ldots$ given by

$$\mathbf{s}(t + 1) = \mathbf{s}(t) + \gamma(t)\mathbf{J}(\mathbf{s})^T \mathbf{P}(\mathbf{x} - \mathbf{w}(\mathbf{s}(t))). \tag{11}$$

Here, $\gamma(t) > 0$ is a step size parameter, $\mathbf{J}(\mathbf{s})$ is the Jacobian $\partial\mathbf{w}(\mathbf{s})/\partial\mathbf{s}$ evaluated at the current point, using eq. (7), and $\mathbf{P}$ a diagonal projection matrix into the subspace $\mathbf{X}^{\text{in}}$, i.e. $P_{ij} = 1$, if $i = j \in I$, and zero otherwise. Equation (11) can be viewed as a *network dynamics* for a recurrent network, however, with node activities represented parametrically by the map coordinate vector $\mathbf{s}$.

In many situations, the simple gradient descent scheme for the determination of the best match point $\mathbf{s}$ turns out to be rather sensitive to the choice of the step size parameter γ. Since the objective function is a quadratic form, a good choice for a minimization algorithm is the Levenberg-Marquard method (see, e.g., [PFTV90]), which automatically adjusts itself from a first order gradient descent far away from the minimum point to a (approximate) second order method in its vicinity. Still, there is a problem of local minima; this cannot be solved in generality, since one can easily devise embeddings of the manifold S and choices of the input vector $\mathbf{x}$ such that two points that are widely separated on S are in the same minimal distance to $\mathbf{x}$. However, one can reduce this problem significantly by taking advantage of the discrete approximation of S that is given by the support points $\mathbf{w_r}$ and taking as a starting value $\mathbf{s}(0)$ for Eq. (11) the "discrete bestmatch vector" $\mathbf{w_s}$ that obeys eq. 10, however, with $\mathbf{s}$ restricted to the discrete set of lattice points $\mathbf{r} \in \tilde{A}$.

5 "Harmonic" PSOMs

The simplest choice for the basis functions $H(\mathbf{a}, \cdot)$ is to make them linear in each coordinate s_i separately. For a d-dimensional space, there are 2^d such polynomials that are independent, and a convenient, orthogonal basis set is given by the 2^d products obtained by choosing for the i-th factor either s_i or $1 - s_i$, $i = 0 \ldots d - 1$. Each such basis polynomial can be considered as "belonging" to a corner $\mathbf{s} = \mathbf{a} \in \{0, 1\}^d$ of the d-dimensional hypercube $[0, 1]^d$ selected by the condition $H(\mathbf{a}, \mathbf{s}) = 1$ and zero on the remaining $2^d - 1$ corners $\mathbf{s}' \neq \mathbf{s}$. Moreover, due to their axis-wise linearity, any function $\phi(\cdot)$ in the space that is spanned by the polynomials H obeys

$$\Delta_\mathbf{s}\phi(\mathbf{s}) = 0,$$

i.e., is *harmonic*. From the viewpoint of constructing approximations, this is a desireable property: the value of a harmonic function at an inner point $\mathbf{s}$ of its domain is always the average of all values of the function on the surface of an (infinitesimal) sphere centered at $\mathbf{s}$. Thus, harmonic functions achieve in some sense the "smoothest" continuation of function values prescribed on the boundary of a domain into its interior (and "soap films" offer a physical realization of this type of interpolation).

In the present case, the boundary is given by the surface of the d-dimensional hypercube, with the possibility to prescribe 2^d independent function values at its corner points (of course, one can choose any other set of 2^d points in "general position", but the above basis functions have been chosen to make the indicated choice the most convenient). The resulting "harmonic PSOMs" share with hyperplanes the property that they vary linearly along each coordinate axis direction; however, the different linear projections now combine to a manifold that is no longer globally linear.

It may be interesting to compare the approximation capabilities of harmonic PSOMs with the linear models. For an intrinsic data dimension of d, a harmonic

PSOM has 2^d free parameters and can pass through the same number of specified points, as compared to $d+1$ parameters and points for a hyperplane. Therefore, for $d > 2$ harmonic PSOMs offer considerably more flexibility to fit to given points than hyperplanes. To illustrate this, we compare the approximation capabilities of a linear model and of a harmonic PSOM for a benchmark function on the d-dimensional hypercube that is essentially the parity function on the 2^d corners of the hypercube, smoothly extended into its interior by forming the sum

$$f(s_1 \ldots s_d) = \frac{\sum\limits_{a \in \{0,1\}^d} [(\sum_i a_i) \bmod 2] e^{-\lambda \|s-a\|}}{\sum\limits_{a \in \{0,1\}^d} e^{-\lambda \|s-a\|}}$$

where parameter $\lambda > 0$ determines how fast the contribution of each corner's parity value $(\sum_i a_i) \bmod 2$ to $f(s)$ decays as s moves away from that corner.

dim d	linear	PSOM (2^d nodes)
2	0.20748	0.04921
3	0.12422	0.03295
4	0.07157	0.01937
5	0.03977	0.01054

Table 1: Root mean square error of least square linear approximation and harmonic PSOM approximation of function $f(\cdot)$ from Eq. 5, based on function values at the 2^d corners of the d-dimensional hypercube $[0, 1]^d$ (resulting approximation error was evaluated on the set of 10^d points of a d-dimensional cubical $10 \times 10 \times \ldots 10$ lattice).

Table 1 shows the average root mean square errors obtained with the linear and the harmonic PSOM, when values for $f(\cdot)$ (with the choice $\lambda = 3$) were given for the 2^d hypercube corners. In each case, the constructed PSOM passes exactly through the given points, while the hyperplane can only be chosen to minimize the mean square error for the given points. In both cases, the error was evaluated on an equidistant 10^d grid superimposed on the unit cube. Figure 4. shows the graph of the function $f()$ for $d = 2$.

6   General PSOMs

The harmonic PSOMs of the previous section are only the simplest realization of PSOMs. While their property of being harmonic functions has some strong virtues from the point of view of function interpolation, the same property also gives rise to an important limitation that they still share with linear models: they cannot represent functions that have local extrema at interior points of their domain. To represent such functions requires to use at least second order polynomials along each coordinate direction, giving rise to a hypercubical lattice of 3^d points on which independent function values may be specified (for an example, cf. the next section 7).

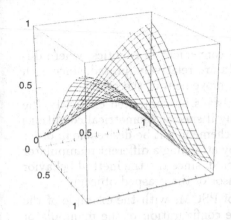

Figure 4: The "smoothed parity function", Eq.(5), for $d = 2$ and $\lambda = 3$.

If there are many support points along an axis direction, the simple Lagrange polynomials become unsuitable due to their rapid growth (this problem however, is less acute than it might seem, since in higher dimensions one usually can only afford a very limited number of support points along each axis direction). One possibility to circumvent this problem has been suggested in [] and consists in making the PSOMs local (this can be viewed as an extension of the LLM-approach to higher order approximations). A different possibility is to stay with polynomials, but to choose a non-uniform spacing of the support values in order to make the higher order polynomials more well-behaved. A particularly suitable choice is a spacing following a cosine law and leading to the Chebyshev polynomials. Refs. [WR95, Wal96] show that both approaches can lead to very significant gains in the approximation accuracy of PSOMs.

Although polynomials provide a convenient set of functions to construct PSOMs, different types of basis functions, such as, e.g., trigonometric functions, can be used as well. The orthogonality requirement Eq.5 is actually only a convenience, but not a strict necessity. If a non-orthogonal function set $\tilde{H}(a, s)$ is chosen, one can easily orthogonalize it by defining as a new set of functions

$$H(\mathbf{a}, \mathbf{s}) = \sum_{\mathbf{a}' \in \tilde{A}} T_{\mathbf{a}, \mathbf{a}'} \tilde{H}(\mathbf{a}', \mathbf{s})$$

where T is the inverse of the matrix M whose elements are $M_{\mathbf{a}, \mathbf{a}'} = \tilde{H}(\mathbf{a}, \mathbf{a}')$. Substituting this in Eq.(7) shows that the use of a non-orthogonal function set thus is equivalent to the use of Eq.(7) with a linearly transformed set of weight vectors $\tilde{\mathbf{w}}$ given by $\tilde{\mathbf{w}}_{\mathbf{a}} = \sum_{\mathbf{a}' \in \tilde{A}} \mathbf{w}_{\mathbf{a}} T_{\mathbf{a}, \mathbf{a}'}$.

One should note, however, that this does not automatically guarantee that the second condition, Eq.(6), is obeyed. If Eq. 6 is violated, a simple translation $\mathbf{w}_{\mathbf{r}} \rightarrow \mathbf{w}_{\mathbf{r}} + \mathbf{t}$ of all data points will no longer entail a corresponding translation $\mathbf{w}(\mathbf{s}) \rightarrow \mathbf{w}(\mathbf{s}) + \mathbf{t}$. While this relationship will remain true for $\mathbf{s} \in \tilde{A}$, it does no longer hold at the intermediate points. As a consequence, a number of desirable properties for an interpolation scheme are lost: e.g., the resulting manifold becomes dependent on the choice of the coordinate origin, and, if the set of given data vectors $\mathbf{w}_{\mathbf{r}}$ lies within a d-dimensional hyperplane, the resulting PSOM will in general exhibit nonlinear "bumps" at the intermediate points.

7 An Example

Many of the properties of PSOMs are particularly convenient in robotics, where different coordinate systems abound and training data are relatively limited, since each training sample usually is associated with an extra movement of the robot.

While the forward kinematics of a robot can always be computed explicitly, the inverse kinematics can only be obtained analytically if special geometrical conditions are satisfied. Even then an adaptive computation scheme might be desireable in order to account for tolerances or changes, e.g., incurred by attaching a different manipulator (this is even more important, if dynamics is considered, since the the inertial behavior of a robot depends on the mass and the inertia tensor of the grasped object).

In the following, we illustrate the capabilities of PSOMs with the example of the inverse kinematics of a six-axis PUMA robot. The configuration of the manipulator of such a robot can be described by six parameters (three positional coordinates and three orientation parameters), but it is frequent practice in robotics to use larger, "redundant" parameter sets that may be more convenient to specify. A conventional set of such parameters are the cartesian coordinates (x, y, z) of the (center of) the manipulator plus six further values to specify the direction (a_x, a_y, a_z) of the hand axis ("approach vector") and the normal (n_x, n_y, n_z) of the manipulator "palm". A different set of parameters are the six joint angles, and the inverse kinematics is the mapping from the first, cartesian coordinate set to the corresponding joint angle values (strictly speaking, for the case of the PUMA robot, this "mapping" has eight different "branches", since – except for singular configurations – each specified cartesian location and orientation of the end-effector can be achieved with eight different joint angle 6-tuples; for a more detailed description, see [FGL87]).

Therefore, the robots configuration space can be viewed as a 6-dimensional manifold S that is embedded in the 15-dimensional space of coordinate parameters $\mathbf{x} = (x, y, z, a_x, a_y, a_z, n_x, n_y, n_z, \theta_1 \ldots \theta_6)$.

Standard learning approaches are concerned with constructing the various mappings that might be required for controlling the robot. The most important two mappings are from the six joint variables $\vec{\theta}$ to the corresponding cartesian parameters $(x, y, z, a_x, a_y, a_z, n_x, n_y, n_z)$ (direct kinematics) and vice versa (inverse kinematics), but "mixed" mappings (specification of some angles and of some cartesian parameters and computation of the unspecified parameters from these) are also conceivable.

While the usual learning approaches require construction of a separate approximation for each of these mappings, PSOMs allow the simultaneous construction of all of them by directly representing the 6-dimensional manifold in its (here 15-dimensional) embedding space.

Construction of the PSOM requires specification of a set of training examples $\mathbf{w_r} = (x, y, z, a_x, a_y, a_z, n_x, n_y, n_z)_\mathbf{a}$, taken for a set of configurations $\mathbf{a}$ that vary over a 6-dimensional cubical "configuration grid" $\tilde{A}$. This grid can be chosen in several ways; for the demonstration here we identified $\tilde{A}$ with the $3^6 = 729$ points of a $3 \times 3 \times \ldots \times 3$ hypercubical grid covering the 6-dimensional "joint angle hypercube" $[\theta_i^0 - 45^0, \theta_i^0 + 45^0]$, $i = 1 \ldots 6$, where θ_i^0 denotes the angle of joint i when the robot is in its "home position". The training points are then obtained by using the (known) forward kinematics of the robot to obtain for each joint angle point the remaining cartesian coordinate values.

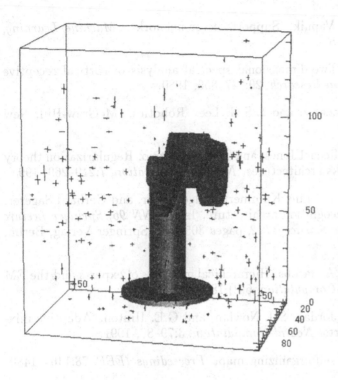

Figure 5: Positioning errors of PUMA robot arm with six degrees of freedom when the inverse kinematics transform is computed with a six-dimensional PSOM with 3^6 nodes, representing the configuration manifold of the manipulator embedded in a 15-dimensional $\vec{r}, \vec{a}, \vec{n}, \vec{\theta}$-space (for details, see text).

Then, the PSOM is constructed according to Eq.(7), with the $\mathbf{w_r}$ given by the 729 15-dimensional vectors obtained in this way.

Fig.5 shows the accuracy of the constructed PSOM by indicating the robot's positioning error with an error cross for a set of 200 randomly chosen target positions in its workspace, using the PSOM for the inverse kinematics computation. The resulting dimensionless NRMS positioning error is 0.06, which compares very favorably with earlier approaches on (simplified versions of) the same task, such as [YB93]. These authors focus only the 3d-task that remains when manipulator orientation is neglected, and they use several thousands of training examples instead of 729.

In the simplified case they consider, the manifold S is only 3-dimensional and we can embed it in a now 6-dimensional space of cartesian coordinates $(x, y, z, \theta_1, \theta_2, \theta_3)$ and joint angles, since orientation is neglected. Staying with the 3 discretization points per manifold dimension now only requires only 27 training examples (as compared to several thousand in [YB93]), leading to a mean root square positioning error of 2.6 cm (corresponding to a NMRSE of 0.05). By using five instead of three nodes per dimension (requiring 125 data samples), the error can be reduced to 0.5cm (NMRSE of 0.0097). By making use of the further improvements considered in [WR95], the error can be reduced even further. For a more recent discussion, including extensions to combining several PSOMs, see, e.g., [WR96a, Wal96].

References

[CV95]   C. Cortes and V. Vapnik. Support-vector networks. *Machine Learning*, 20(3):273–297, 1995.

[Dau80]   J. G. Daugman. Two-dimensional spectral analysis of cortical receptive field profiles. *Vision Research*, 20:847–856, 1980.

[FGL87]   K.S. Fu, R.C. Gonzalez, and C.S.G. Lee. Robotics. McGraw-Hill, New York, 1987.

[GJP95]   Frederico Girosi, Michael Jones, and Tomaso Poggio. Regularization theory and neural networks architectures. *Neural Computation*, 7:219–269, 1995.

[HKRS96]   Gunther Heidemann, Franz Kummert, Helge Ritter, and Gerhard Sagerer. A hybrid object recognition architecture. In *ICANN 96, Springer Lecture Notes in Computer Science 1112*, pages 305–310. Springer Verlag, Berlin, 1996.

[JJ94]   M.I. Jordan and R.A. Jacobs. Hierarchical mixtures of experts and the EM algorithm. *Neural Computation*, 6(2):181–214, 1994.

[JJNH91]   R.A. Jacobs, M.I. Jordan, S.J. Nowlan, and G.E. Hinton. Adaptive mixtures of local experts. *Neural Computation*, 3:79–87, 1991.

[Koh90]   T. Kohonen. The self-organizing map. *Proceedings IEEE*, 78:1464–1480, 1990.

[LR97]   Enno Littmann and Helge Ritter. Adaptive color segmentation – a comparison of neural and statistical methods. *IEEE Transactions on Neural Networks*, pages 175–185, 1997.

[MD88]   J. Moody and C. Darken. Learning with localized receptive fields. In *Connectionist Models: Proceedings of the 1988 Summer School*, pages 133–143. Morgan Kaufman Publishers, San Mateo, CA, 1988.

[MR92]   Andrea Meyering and Helge Ritter. Learning 3d-shape-perception with local linear maps. In *Proceedings of the International Joint Conference on Neural Networks*, pages IV:432–436, Baltimore, 1992.

[PFTV90]   W.H. Press, B.P. Flannery, S.A. Teukolsky, and W.T. Vetterling. *Numerical Recipes in C*. Cambridge University Press, Cambridge, 1990.

[Rit93]   Helge Ritter. Parametrized self-organizing maps. In S. Gielen and B. Kappen, editors, *ICANN 93-Proceedings*, pages 568–577, Berlin, 1993. Springer.

[RMS92]   Helge Ritter, Thomas Martinetz, and Klaus Schulten. *Neural Computation and Self-organizing Maps*. Addison Wesley, [1. revised english edition] edition, 1992.

[RS87]   Helge Ritter and Klaus Schulten. Extending kohonen's self-organizing map-
           ping algorithm to learn ballistic movements. In Eckmiller R. and von der
           Malsburg C, editors, *Neural Computers*, pages 393–406, Heidelberg, 1987.
           Springer.

[SU90]   K. Stokbro and D.K. Umberger. Forecasting with weighted maps. In Martin
           Casdagli and Stephen Eubank, editors, *Nonlinear Modeling and Forecast-
           ing*, volume 4, pages 73–98. Addison-Wesley, 1990.

[Wal96]   Jörg Walter. *Rapid Learning in Robotics*. Cuvillier Verlag Göttingen, 1996.
           also http://www.techfak.uni-bielefeld.de/~walter/pub/.

[WR95]   Jörg Walter and Helge Ritter. Local PSOMs and Chebyshev PSOMs –
           improving the parametrised self-organizing maps. In *Proc. Int. Conf. on
           Artificial Neural Networks (ICANN-95), Paris*, volume 1, pages 95–102,
           1995.

[WR96a]   Jörg Walter and Helge Ritter. Investment learning with hierarchical PSOM.
           In D. Touretzky, M. Mozer, and M. Hasselmo, editors, *Advances in Neu-
           ral Information Processing Systems 8 (NIPS*95)*, pages 570–576. Bradford
           MIT Press, 1996.

[WR96b]   Jörg Walter and Helge Ritter. Rapid learning with parametrized self-
           organizing maps. *Neurocomputing*, pages 131–153, 1996.

[YB93]   D. Yeung and G. Bekey. *on reducing learning time in context dependent
           mappings. IEEE Transactions on Neural Networks, 4:31–42, 1993.*

STATISTICS AND NETWORKS

ADAPTIVE MARKET SIMULATION AND RISK ASSESSMENT

R.A. Müller

Daimler-Benz AG, Berlin Germany

ABSTRACT

Quantitative Reasoning (QR) is a powerful method for simulation and risk assessment. QR and applications are presented, allowing the representation of market and company structures (demand and supply side, prices, values, resources, budgets, ...), dynamics (growth, structural changes, ...), interdependencies (e. g. between market and company related factors). The appoach provides the following features to end users:

- calculating, simulating without programming,
- revealing hidden side effects, if-then analysis,
- checking complex plans and forecasts for consistency,
- accounting for risks and uncertainties (calculating with interval numbers),
- incorporating rich and heterogeneous market expertise,
- providing reliable results,
- adaptivity related to new information (Bayesian inference),
- supporting cooperative work.
- One of the most important examples constructed with QR currently in practice is a planning support tool called GVE. It was built by Daimler-Benz Research for the Mercedes-Benz (MB) marketing department. GVE, the German acronym for the European Commercial Transportation Market, models MB's business opportunities and risks as determined by the behavior and structural changes of the overall market.

1. INTRODUCTION

Theories concerning vagueness and uncertainty have a long tradition in science and engineering. This is well documented by the development of probability theory and statistics. Probability theory for instance has played a key role in the development of such important engineering fields as communication and control (see e.g. [1], [2]). Within the AI community it was recognized very early that the simulation of human problem solving capability requires concepts for the representation of uncertainty. This lead to the (re)discovery, (re)invention and (re)development of numerous techniques ranging from nonclassical or modal logic to probability and measure theory with more or less successful attempts to integrate them with existing knowledge representation concepts (see [3], [4] for a survey). Some of the well known references are Zadeh's "fuzzy set" theory [5] or the Dempster-Shafer theory of evidence [6]; see also [7], [8], [9], [10].

There have been several attempts to apply control engineering methods for economic systems modeling [11], [12]. Whereas within engineering applications dealing with technical systems (e.g. power engines) it is in most cases possible and adequate as well to build models describing the systems dynamic behavior, this is unfortunately not the case for social systems (e.g. markets). Social systems have a much higher degree of unpredictability than technical systems have. Social systems lack the laws of nature engineers successfully use for accurate prediction and control of technical systems. Therefore, model based prediction of social systems and describing the proper rules of state transformation is not only gradually different from technical system modeling but fundamentally different.

The main reason why the above mentioned methods failed was the usage of inadequate models; model variables with previously physical interpretation (like power, pressure, voltage) have just been reinterpreted within a socioeconomical context, and - moreover - the same kind of models (successfully) used for describing physical interdependencies were attempted to describe the dynamics of social systems. The role strong laws of nature play within technical system models was replaced by weak (and naive) assumptions about social systems behavior. A model like Ohms law for instance, describing the relationship between current and voltage within an electrical system, even in the form of a nonlinear and sto-chastic differential equation is inadequate (with respect to its predictive power) to describe dependencies between wages and prices, because there is a fundamentally different kind of relationship between wages and prices than between current and voltage: whereas those physical dependencies are independent from human actors, this apparently does not hold true for social systems whose main 'components' are human beings with permanently changing behavior and behavioral patterns respectively. If, for instance an adequate general stock market forecasting model would be used by the market participants, than its use would immediately change the basic behavior of the market, and the behavioral patterns of the participants.

In order to overcome this situation, Schmid has proposed an entirely different approach [13], namely to use just an other category of models: instead of modeling the system dynamics of the 'world', to model the information we have about the respective system under consideration. Although the QR modeling approach to be presented in the next chapter makes full use of the above mentioned state space modeling framework used by control system engineers, it has to be kept in mind that the variables within QR models have a fundamentally different interpretation than those of traditional state space modeling: QR models are semantical models describing the state of knowledge we have about a certain system, not the states of the system itself.

The concept of Quantitative Reasoning is based on Schmid's work [13]. This paper will describe the QR approach with special reference to market simulation (scenario based forecasting and risk evaluation) and to models and experiences made at Daimler-Benz within the last years. The application situation will be described in the next chapter.

2. APPLICATION BACKGROUND

The marketing department of a company wants to determine the sales figures of future products two to five years ahead. For the sake of simplicity we will assume that the products are completely defined in advance. There are basically two approaches the forecaster may follow in order to reach the goal:

- he could rely on a given model describing the market dynamics, for instance as a high order differential equation system, identifying the parameters and calculating the required forecasting figures, or
- collecting all available information within the special context of his forecasting task, and combining them into a coherent description of future market states.

The advantage of the first approach, once the type of the mathematical model and enough data from past periods are given, is its simplicity to calculate the forecasts. But their reliability depends completely upon the chosen type of model, and the validity of the latter in general can never be guaranteed in advance.

The second approach suffers at first from the abundance of information. There is no appropriate methodology to combine very heterogeneous information (like texts and numerical values) into a coherent framework, and to transform such redundant raw information from many sources into the required market forecasts.

Both approaches have in common the problem of observation errors: there is no measurement, no observation without error. Moreover, the quality of different information may be very different, dependent for instance upon the sources of information.

Business managers and marketing professionals however would prefer the second approach, since only the latter would allow to eliminate the arbitrariness of any chosen mathematical model. They would also require to know the reasons for the forecasts. Extrapolation, just another arbitrary and purely mathematical (no logical) operation is only plausible if a person believes in trends, it never explains a trend.

Quantitative Reasoning follows the second approach. It avoids the arbitrariness of

- mathematical models in general (using semantical models instead), and
- extrapolation as a method in particular (using logical inference instead)

3. QUANTITATIVE REASONING (QR)

Quantitative Reasoning (QR) is a computerized methodology for quantitative modeling and simulation, especially accounting for uncertainty. Its most important special feature is the modeling approach. Models are always semantical models; they are equivalent to definitorial equations within mathematical simulation models as used by engineers. Only the semantical relationships between variables are represented, nothing else. This feature allows a business professional without major mathematical skills always to clearly understand and modify such a model. The values of the model variables are interval numbers (which are interpreted as confidence intervals). The user of a model does not need to perform any procedural kind of programming.

QR's calculation is statistical (Bayesian) inference driven by the meaning of the variables within the semantical model; it is a kind of generalized and "intelligent" arithmetic which differs from its ordinary form in the following way:

- arithmetic operates on interval numbers (which are interpreted as confidence intervals),
- there is no procedural definition of operations, but only declarative (very similar to knowledge representation in Artificial Intelligence).

Another valid definition of QR could be a propagation method for quantitative constraints related to semantical networks. Although the modeling framework of QR consisting of

- an object language definition (structured variable list), and
- linear (+) and nonlinear (:) relations between these list elements

at first glance seems rather primitive, it allows for the representation of surprisingly rich and complex structures. This ability is based on the following modeling features

a)   tree structured objects,
b)   chaining of relations,
c)   "packaging" of quotients as matrices.

The above mentioned representation of tree structures has a built-in attribute inheritance mechanism: if the class of objects (say cars) is divided by a first attribute "price" into the class "luxury" and "economy" and by a second attribute "fuel consumption" into "high" and "low", then the model "knows" that there are for instance both high and low consumption luxury cars

3.1 Semantical Modeling

A model consists of a structured list of object variables connected by relational variables constituting a semantical network. Each object variable is a vector x referring to real world objects, which have always to be countable or measurable. The classes x_i represented by these object variables may be partitioned into subclasses by arbitrary tree structures to be defined by attribute hierarchies, e.g. commercial vehicle demand by (type of car $\otimes$ type of goods $\otimes$ type of traffic $\otimes$ ...), where '$\otimes$' denotes the Cartesian product between the set of attribute values for each of the types respectively.

Between object variables two types of relations, linear combinations and quotients may be defined. The general form of a linear combination is a scalar product

$$y: = a^*x$$

of two vectors $a = (a_1, a_2, ... a_n)$ and $x = (x_1, x_2, ... x_n)$, where y and the x_i's are members of the object list given above and the a_i's are arbitrary real or integer numbers. Each x_i may be the left hand side of another linear combination. A quotient is a scalar relation of the type

$$y: = x_i : x_j$$

where y, x_i and x_j are members of the object list. In contrast to linear combinations the right hand side is not allowed to contain quotients.

Although at first glance the modeling framework seems rather simple, it allows for the representation of surprisingly rich and complex real world structures (e.g. time varying state by using quotients of the type $y := x_{i,t+1}/x_{i,t}$). The QR framework automatically generates matrix structures for y depending on the structures of object vectors (e.g. state transition matrix). A product relation does not need to be defined explicitly since the quotient relation implies the product definition $x_i := y*x_j$. The following figure gives an example of a simple financial model without any class partition, consisting of three object variables Turnover, Cost, Capital, one linear combination Profit := Turnover - Cost; ($a_1 =$ 1 and $a_2 = -1$), and three quotients (e.g. Return on Investment (ROI) := Profit : Capital).

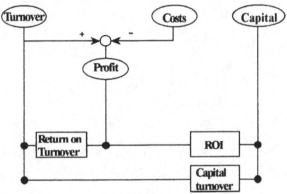

Figure 1:   Semantical network example (Dupont business indicators)

It is important to note that both linear combination and quotients are pure definitions, and do neither entail any arithmetic procedure nor any assumption about the state of the world the model refers to or its changes. Models built in this way are easy to validate by only checking the validity of these definitions. The semantical modeling level is strictly separated from the empirical part containing the state description consisting of assumptions, observations, and measurement results.

3.2   State Description and risk evaluation

A state description is given by assigning numerical data (e.g. measurement results) to each of the variables within the semantical network. A range of uncertainty has always to be given in brackets behind the number. The expression "$x = a$ (b)" means "the value of the variable x lies between the lower limit a - b and the upper limit a + b", and there is a risk of, say 1% probability for the value of x exceeding these limits. The interval is interpreted as range of confidence based on a given probability measure, e.g. "price = 10 (2) $" means that given a 99% confidence level the true value lies between 8 and 12 $. The bandwidth may be interpreted both as measurement error or range of fluctuation. A zero bandwidth is also possible as a special case. By successively putting one data element after the other into the network the inference process is able to calculate missing values by reducing the uncertainty (bandwidth) of given values without any procedural programming by the user.

There is no fixed and predetermined distinction between input and output variables. Any variable generally is input and output as well. The mechanism of forward and backward computation is only controlled by the meaning (network structure and bandwidth) of the variables. There is no need for a user defined procedure.

Within the QR tool there is a strict separation between the structural level of representation, the data level and the inference machine (for transforming and estimating data). This kind of architecture and information processing is very similar to rule based or constraint propagation methods of knowledge based approaches. Such architectures are of great advantage since they are independent of any application: though QR is used in very different applications, its inference machine has never to be changed.

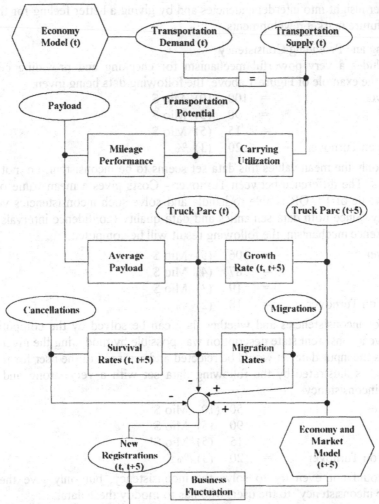

Figure 2:   Commercial Vehicle Demand Model (simplified version)

Each variable represents a constraint to each other variable, which is propagated through the network by using the Bayesian rules for conditional probabilities. Since, in general, different values depending on different propagation paths will be calculated for the same variable, contradictions will arise as a rule. QR includes a powerful mechanism for checking and providing consistency. Hidden side effects become transparent to the user. Different scenarios based on different assumptions concerning both the market and the risk of changes are calculated and compared. Market forecasts from other sources are checked for consistency. Own evaluations are proved or disproved. The risk of producing e.g. a certain number of vehicles of a special kind (new registrations) can be traced back to the risks of economic development, traffic structure, vehicle properties and other variables shown in Figure 2. The tool provides an improved foundation of management decisions by giving a better insight into interdependencies and by giving a better feeling for the chances and risks of future market developments.

3.3 Checking and Providing Consistency

QR includes a very powerful mechanism for checking and providing consistency. Referring to the example of Figure 1 above, the following data being given:

Turnover	=	100	(10)	Mio $	
Costs	=	90	(5)	Mio $	
Profit	=	15	(5)	Mio $	
Return on Turnover	=	20	(3)	%.	

Considering only the mean values this data set seems to be inconsistent, i.e. not free from contradictions. The difference between Turnover - Costs gives a mean value of 10 Mio, and not the value of 15. QR is able to handle and solve such inconsistencies very simply and intuitively. Since both data semantics and data quality (confidence intervals) are used by QR's inference mechanism, the following result will be computed:

Turnover	=	106	(5)	Mio $	
Costs	=	87	(4)	Mio $	
Profit	=	19	(3)	Mio $	
Return on Turnover	=	18	(2)	%.	

QR checks for inconsistencies and whether they can be solved by the computer. In this example above a consistent state description was possible by modifying the given intervals. In other cases the input data set would be rejected and returned to the user for further data modifications, as illustrated by the following data set with a very strong and internally unresolvable inconsistency,

Turnover	=	50	(10)	Mio $	
Costs	=	90	(5)	Mio $	
Profit	=	15	(5)	Mio $	
Return on Turnover	=	20	(3)	%.	

Here, QR would not even try to solve the inconsistency, but only give the message "unresolvable inconsistency" to the user, who has to modify these data.

4. APPLICATIONS

Several prototypes and working systems have been realized (e.g. [14], [15], [16], [17]). The example of figure # above is a very simplified version of the European transportation and vehicle demand market model as used by the Mercedes-Benz marketing department for strategic sales planning and forecasting. The next figure shows a small part of EXBUS, a system supporting the consultancy of public transport authorities [17], developed also at Daimler-Benz Research Berlin (figure 3).

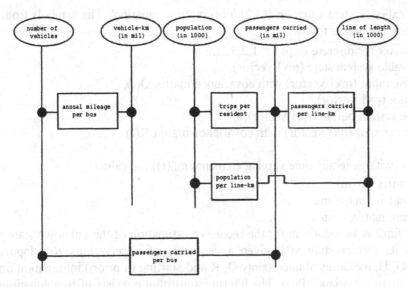

Figure 3: part of EXBUS semantical net

5. UNDERLYING THEORY AND CONCEPTS

QR is a combination of several concepts:

- state space models
- stochastic filtering
- logic of sorts/classes and semantical net representation.

It's most important cornerstone is the state space concept as founded by Kalman, expanded and popularized by Zadeh and Desoer (1963), which still is the theoretical backbone of modern control engineering. It was Kalman who developed a strong and far reaching theory for recursive state estimation, which is well known as Kalman filtering [1]. QR's inference mechanism is an extended version of the nonlinear Kalman filter developed by Jazwinsky (1970). There have been many attempts to widen the scope of Kalman filtering outside the control engineering field (e.g. [11]; [12]).

One of the most interesting extensions was Schmid's approach to combine the state space modeling and Kalman filtering framework with concepts of sort logic and semantical nets as used by AI, which he called "accounting models" or "balance models" ([13]; [19]). The standard (linear) state space model with the Kalman filter algorithm is given below.

State Space Model:

$$x(t+1) = A(t)x(t) + G(t)u(t) + w(t) \qquad (5.1)$$

$$z(t) = H(t)x(t) + v(t) \qquad (5.2)$$

Eq. (5.1) is called 'system equation', (5.2) 'observation equation'. The symbols from eqns. 5.1 and 5.2 are defined as follows:

t: time parameter for discrete steps $t = 1,2,3,...$,
x: not observable system state (mx1 vector),
w: white noise input (mx1 vector) with covariance matrix $Q(t)$,
u: system input (px1 vector),
z: observable system output (qx1 vector),
v: white noise output (qx1 vector) with covariance matrix $R(t)$.

The matrices (with generally time varying elements $m_{ij}(t)$) are called
A: System matrix (mxm),
G: Control matrix (mxp) and
H: Observation matrix (qxm).
The Kalman filter is an algorithm for the recursive estimation of the unknown state vector $x(t+1)$ from its former state $x(t)$, given a sequence of observations $z(t)$, inputs $u(t)$, relations A, G, H, measures of uncertainty Q, R and starting (a priori) information on $x(0)$, including its error covariance $P(0)$. The filtering algorithm consists of two equations, one for the estimation of the mean of x (eq. 5.3), the other for the estimation of the confidence intervals and correlation of the error represented by the covariance matrix P (eq. 5.4). The algorithm evaluates the (a-posteriori) conditional probability density of the random variable $x(t+1)$ given all relational information (matrices A,G,H,P,Q,R), inputs u and observations z up to time $t+1$, and a-priori information on x up to time t.

Kalman filter

$$x(t+1) \quad = \quad xx(t) + K(t+1)(z(t+1) - H(t+1)xx(t)) \tag{5.3}$$

$$xx(t) \quad = \quad A(t)x(t) + G(t)u(t) \tag{5.3a}$$

$$P(t+1) \quad = \quad (I - K(t+1)H(t+1))PP(t) \tag{5.4}$$

with A' denoting the transposed of A, I the unit matrix, and the 'gain' matrix K being

$$K(t+1) \quad = \quad PP(t)H(t+1)'(H(t+1)PP(t)H(t+1)' + R(t+1))-1 \tag{5.4a}$$

$$PP(t) \quad = \quad A(t)P(t)A(t)' + Q(t) \tag{5.4b}$$

These equations constitute the computational kernel as one part of the inference machine. The other part is a compiler translating the user defined model as described in ch. 2.1 (user terminology, semantical relations, empirical data and uncertainty/quality) into the internal state space representation (eq. 5.1, 5.2). The filtering equations (5.3, 5.4) are generated automatically during the run time depending on the current situation (which data are given, which are required). Schmid's effort has transformed the epistemic status of Kalman filtering from a purely statistical framework into a logical framework with truth preserving capabilities [20]. QR can be considered as a special case of constraint based reasoning. It controls the propagation of quantitative constraints (interval scaled numbers) through a semantical network, where the network is defined by the users object language.

REFERENCES

1. Kalman, R.E.: A New Approach to Linear Filtering and Prediction Problems, in: Journal of Basic Engineering, March 1960, 35-45.
2. Jazwinsky, A.H.: Stochastic Processes and Filtering Theory, Academic Press, New York 1970.
3. Kruse, R. Siegel,P. (eds.): Symbolic and Quantitative Approaches to Uncertainty. Proceedings of the European Conference ECSQAU, Marseille, France, October 15-17, 1991, Berlin (Springer) 1991.
4. Kruse, R. Gebhardt, J. Klawonn, F.: Modellierung von Vagheit und Unsicherheit, Fuzzy Logik und andere Kalküle, in: Künstliche Intelligenz 4, 1991, 13-17.
5. Zadeh, L.: Fuzzy Sets. Information and Control. 8, 338-353, 1965.
6. Shafer, G.: A Mathematical Theory of Evidence. Princeton NJ (Princeton University Press) 1976.
7. Skala, H.J. Termini, S. Trillas, E. (eds.): Aspects of Vagueness. Dordrecht (Reidel) 1984.
8. Goodman, I.R. Nguyen, H.T.: Uncertainty Models for Knowledge-Based Systems, A Unified Approach to the Measurement of Uncertainty, Amsterdam (North Holland) 1985.
9. Kanal, L.N. Lemmer, J.F. (eds.): Uncertainty in Artificial Intelligence, Amsterdam (North Holland) 1986.
10. Furstenberg, G. von: Acting under Uncertainty, Multidisciplinary Conceptions, Boston (Kluwer) 1990.
11. Aoki,M.: Optimal Control and System Theory, in: Dynamic Economic Analysis, New York 1976.
12. Harrison,P.J. Stevens, C.F.: Bayesian Forecasting. J.Roy.Stat.Soc. B 38, 1976, 205-247.
13. Schmid, B.: Bilanzmodelle, Simulationsverfahren zur Verarbeitung unscharfer Teilinformationen. Bericht des ORL-Instituts der ETH Nr. 40, Zürich 1979.
14. Huber, H. Schmid, B.: Perspektiven des Energiewesens in der Schweiz und räumliche Konsequenzen, Zürich 1984.
15. Kyburz, R. Schmid, B.: Accounting Models - A New Tool in Forecasting, Studienunterlagen zur Orts-, Regional- und Landesplanung Nr.40. ORL-Institut, ETH Zürich 1979.
16. Müller, R.A. Reske, J. Minx, E.P.W.: Eine Zukunftsanalyse der Ausstattung privater Haushalte mit PKW in der Bundesrepublik Deutschland bis zum Jahre 2010, in: Daimler-Benz-AG (Hrsg.): Langfristprognosen - Zahlenspielerei oder Hilfsmittel für die Planung? Referate des 6.Daimler-Benz-Seminars der Forschungsgruppe Berlin vom 15. und 16.11.1984. Report Nr.5 der Schriftenreihe der DBAG. Düsseldorf (VDI-Verlag) 1985.

17. Leuthardt, H. Günther, R.: Mit dem PC auf der Spur von Schwachstellen im Verkehrsbetrieb. Rechnergestütztes Verfahren zur Wirtschaftlichkeitsuntersuchung, in: Der Nahverkehr, Heft 3 1989.
18. Zadeh, L. Desoer,C.A.: Linear System Theory. New York (McGraw Hill) 1963.
19. Schmid, B.(ed.): Information on Complex Systems - Representation and Inference. Five Papers Presented at the 4th International Symposium on Forecasting, London 1984. Zürich (Verlag der Fachvereine) 1988.
20. Schmid, B.: Darstellung und automatische Auswertung quantitativer Information. Arbeitsbericht des Instituts für Wirtschaftsinformatik (IWI) der Hochschule St. Gallen. St. Gallen 1992.

PROCESSING OF PRIOR-INFORMATION IN STATISTICS BY PROJECTIONS ON CONVEX CONES

E. Rödel

Humboldt University Berlin, Berlin, Germany

ABSTRACT.

We investigate a Random-Search-Algorithm for finding the projection on a closed convex cone in R^p with respect to a norm defined by any positive definite matrix. It is shown that this algorithm converges almost surely. The power of the algorithm is demonstrated by examples from statistics in which processing of prior information may be formulated as projections of parametervectors on polyhedral cones.

Key words. Random search, projection on a convex cone, optimization in statistics, statistical estimation under prior information.

1. INTRODUCTION AND MOTIVATION

We consider the norm

$$\|y\|_C^2 = y^T C y$$

in R^p for any positive definite matrix C. $\|\ \ \|$ denotes the Euclidian norm.
Let K be a closed convex cone in R^p. There exists a unique $y^* \in K$ to each $y \in R^p$ such that

$$r^* = \inf_{x \in K} \|y - x\|_C = \|y - y^*\|_C. \tag{1}$$

y^* is called the projection of y on K.
Projections on closed convex sets are widely applied in the fields of approximation theory, stochastic optimization and especially of statistical inference, e.g. for modifying estimators under special hypotheses. Such hypotheses may be certain orders among the components of a mean vector, certain assumptions about the type of a regression or a density function. The numerical solution of a projection may be obtained by methods of quadratic optimization or by special algorithms, e.g. the "Pool-Adjacent-Violators-Algorithm" for the estimation of ordered means, see [1]. The quadratic optimization approach leads to the use of well-known algorithms and corresponding programs, which are generally not tailored for the specific demands of projections on convex sets. For this reason, we will propose a simple numerical procedure for finding such projections.

2. A RANDOM SEARCH PROCEDURE FOR PROJECTION ON CONVEX CONES

Obviously, if $y \in K$ then we have $y^* = y$ and consequently, the case $y \notin K$ is only of interest. We propose the following algorithm for finding y^*. Let $C_r(x)$ be the sphere with centre x and radius r with respect to the norm $\|\ \|_C$.

step 0 $n := 0$,
         Find any $\gamma_n \in Int(K)$ and calculate $\rho_n = \|\gamma_n - y\|_C$!
step 1   Generate a random $\tilde{\gamma}_n$ uniformly distributed over
         $\{x : x = \gamma_n + \lambda(y - \gamma_n), \lambda \geq 0\} \cap K_n; K_n = K \cap C_{\rho_n}(y)$!
step2   Generate a random vector φ_n uniformly distributed over the Euclidian unit sphere, i.e. $\|\varphi_n\| = 1$,
and a random vector ζ uniformly distributed over
         $\{x : x = \tilde{\gamma}_n + \lambda\varphi_n, \lambda \geq 0\} \cap K_n$,
         $n := n+1, \gamma_n := \zeta, \rho_n := \|\gamma_n - y\|_C$!
step3   If stop-criterion (see section 4) then finish, else goto step 1 !

We can describe these steps in the following way. Starting from γ_n we go in the direction of y until the border of K and reach the point β_n (see Fig. 1). Now we generate $\tilde{\gamma}_n$ uniformly distributed on the segment $[\gamma_n, \beta_n]$. At the end we go from $\tilde{\gamma}_n$ along the random direction φ_n until the border point $\tilde{\beta}_n$ of K_n and generate γ_{n+1} uniformly distributed on the segment $[\tilde{\gamma}_n, \tilde{\beta}_n]$.

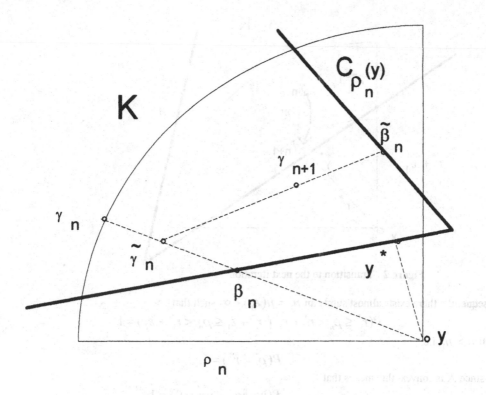

Figure 1 : A scheme of the algorithm

3. CONVERGENCE

It is easy to see that the sequence $\{\rho_n\}$ of distances between y and γ_n is almost surely decreasing, i.e.

$$P(\rho_n < r | \rho_{n-1} = r) = 1.$$

Furthermore, we have

$$P(r^* \leq \rho_n \leq r_0 | \rho_0 = r_0) = 1 \quad \forall n,$$

where r^* is the distance of y to K defined by (1), and consequently, $\{\rho_n\}$ is a bounded positive super-martingale. There exists a random variable ρ^* by the martingale convergence theorem (e.g., see [3]) such that

$$P(\lim_{n\to\infty} \rho_n = \rho^*) = 1$$

The construction of the algorithm implies that for all $0 < \varepsilon_1 < \varepsilon_2$ and all $n \geq 0$

$$P(r^* \leq \rho_{n+1} < r^* + \varepsilon_1 | r^* + \varepsilon_1 \leq \rho_n < r^* + \varepsilon_2) > 0,$$

see figure 2.

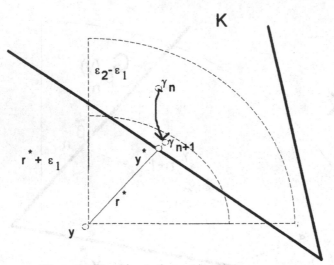

Figure 2 : Transition to the next iteration

Consequently, there exists almost surely an $n_1 = n(\varepsilon_1) < \infty$ such that
$$P(r^* \leq \rho_n < r^* + \varepsilon_1 \mid r^* + \varepsilon_1 \leq \rho_{n_1} < r^* + \varepsilon_2) = 1$$
for all $n > n_1$, i.e.
$$P(\rho^* = r^*) = 1.$$
But, since K is convex, this means that
$$P(\lim_{n \to \infty} \|\gamma_n - y^*\| = 0) = 1.$$

The algorithm is a so-called Pure Adaptive Search algorithm (PAS), i.e. it forces improvement in each iteration. It is known that the expected number of necessary PAS-iterations grows at most linearly in the dimensions of the problem (see [5]).
Another scheme could be the following(see figure 3).

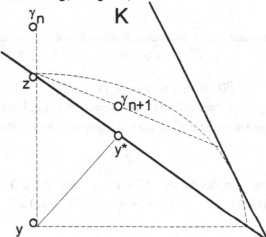

Figure 3 : An alternative scheme.

We assume that K is a polyhedral cone, i.e.

$$K = \{x \in R^p : Ax \le 0\},$$

where A is a (m,p) – Matrix.
The border point z is determined in the following way.

$$z = \gamma_n + \lambda^*(y - \gamma_n),$$

where

$$\lambda^* = \min\{\lambda_i : \lambda_i > 0; i = 1.2, \ldots, m\}, \lambda_i = \frac{-a_i \gamma_n}{a_i(y - \gamma_n)}$$

and a_i is the i-th row of A.
The problem of this iteration schedule consists in the efficient generation of a random direction going from the border point z into the interior of the intersection of K and the ball $C_{\|z-y\|}(y)$, or, equivalently, to find an inner point of this intersection. For this reason we propose the following method.
At first we project y on the line

$$L = \{x : x = \alpha \cdot z; \alpha \ge 0\},$$

i.e. we determine $z^* = \alpha^* \cdot z$, where

$$\alpha^* = \begin{cases} \dfrac{z'Cy}{y'Cy}, & \text{if} \quad z'Cy > 0 \\[2ex] 0 & \text{else} \end{cases}$$

z^* is the nearest point to y in the intersection of K and L. The next iteration γ_{n+1} may be taken as a random point on the intersection $C_{\|z-y\|}(y) \cap [\gamma_n, z^*]$.

4. NUMERICAL IMPLEMENTATION

The algorithm has been implemented for polyhedral cones K on IBM-compatible PC-s in PASCAL and on workstations under UNIX in C.
Numerical difficulties may occur by determining the intersection of a random direction starting from an inner point of K_n with the border of K_n. The feasible region may be left by rounding errors and artificial corrections become necessary such as projections on hyperplanes or seeking an "almost border point" along a direction. The algorithm tends also to stagnate in small corners of K_n and must be supported in such cases by similar corrections as mentioned above. Unfortunately, such actions may slow the convergence considerably. We stop the algorithm at the first point which does not change in the course of a (large) number (100 or more) of iterations.
The advantage of PAS consists in the small storage requirement compared with usual procedures for quadratic optimization, e.g. the procedures of Hildreth-d'Esopo or Powell-Fletcher , see [2]. In contrast to PAS these procedures need arrays of size m^2, where m is the number of nonretundant linear inequalities defining K.

5. APPLICATIONS IN STATISTICS

Let us assume that y is any estimator, not necessarily the Maximum Likelihood Estimator, of an unknown parameter vector θ and that we have the prior information $\theta \in K$, where K is a polyhedral cone defined by

$$K = \left\{ x \in R^p : Ax \leq 0, A = A(m,p) \right\}. \tag{2}$$

We seek the estimator $y^* \in K$ modified by the projection of y on K using a norm $\| \ \|_C$ defined by a positive definite matrix C fitted to the special estimation problem. We will use the notation $r_{pas}^* = \left\| y - y_{pas}^* \right\|_C^2$, where y_{pas}^* is the approximation of y^* found by PAS.

5.1 ISOTONIZATION OF MEANS

Let $y = (y_1, \ldots, y_p)'$ be a vector of observed arithmetical means based on samples of sizes n_i $(i = 1, \ldots, p)$ from p independent populations with true means $\mu_i (i = 1, \ldots, p)$ and a common standard deviation σ.
For testing

$$H_0 : \mu_1 = \mu_2 = \ldots = \mu_p \text{ against } H_K : \mu_1 \leq \mu_2 \leq \ldots \leq \mu_p$$

we need the projection $y^* = (y_1^*, \ldots y_p^*)'$ of y on the polyhedral cone (2) defined by $A = (a_{ij})_{i=1,\ldots,p-1}^{j=1,\ldots,p}$, $a_{ij} = 1$ if $i = j$, $a_{ij} = -1$ if $j = i+1$ and $a_{ij} = 0$ else; $C = diag(w_1, \ldots, w_p)$, $w_i = n_i \sigma^{-2}$.
An efficient solution is given by the "Pool-Adjacent-Violators-Algorithm" (PAVA, see [1]). In the following example we will compare this method with PAS.

Example 1. $\sigma^2 = 10, p = 4$; $y = (15.0 \ \ 13.4 \ \ 17.2 \ \ 16.4)$; $w = (0.5 \ 1.2 \ 1 \ 1.4)$.
The PAVA yields $y_1^* = y_2^* = 13.871, y_3^* = y_4^* = 16.733, r^* = 1.277$. PAS
yields $y_{1(pas)}^* = y_{2(pas)}^* = 13.852, y_{3(pas)}^* = y_{4(pas)}^* = 16.776, r_{pas}^* = 1.2815$ after 158 iterations.
Of course, the PAVA is more efficient than PAS, but the accuracy of PAS is sufficient.

5.2 ISOTONIZATION OF POLYNOMIAL REGRESSION

We consider the regression model

$$Y_i = \sum_{j=1}^p \theta_j t_i^{j-1} + \varepsilon_i, E\varepsilon_i = 0, E(\varepsilon_i \varepsilon_k) = \delta_{ik}\sigma^2, t_i \in [a,b], i,k = 1, \ldots, n.$$

We assume that the design matrix

$$X = (t_i^{j-1})_{i=1,\ldots,n; j=1,\ldots,p}$$

has full rank. Let $\hat{\theta} = (\hat{\theta}_1, \ldots, \hat{\theta}_p)'$ be the LSE of $\theta = (\theta_1, \ldots, \theta_p)'$. Furthermore, let us assume that the true polynomial regression is nondecreasing on [a,b], i.e.

$$\theta \in \tilde{K} = \left\{ \theta \in R^p : \sum_{j=2}^p (j-1)\theta_j t^{j-2} \geq 0 \ \ \forall t \in [a,b] \right\}.$$

$\tilde{K}$ is a closed convex cone defined by a continuum of linear restrictions on θ. We will replace $\tilde{K}$ by K in the following way :

$$K = \left\{ \theta \in R^p : \sum_{j=2}^{p} (j-1)\theta_j u_i^{j-2} \geq 0, u_i \in [a,b], l = 1,...,N \right\},$$

where $\{u_l\}$ is a sufficiently fine grid on [a,b]. With

$$A = (a_{lj})_{l=1,...,N;j=1,...,p}, a_{lj} = -(j-1)u_i^{j-2}$$

we get for K the standard representation (9). As usually, we set $C = X'X$.

Example 2. Let
$$P(t) = 1 + 2t - 2t^2 + 0.5t^3 + t^4$$
be the true regression function in the model
$$Y_i = P(t) + \varepsilon_i \ (i = 1,...,100),$$

$$p = 5, \quad a = -0.5 \leq t_i = -0.5 + (i-1)/99 \leq 0.5 = b \quad (i = 1,...,100),$$

$$\varepsilon_i \sim N(0, \sigma^2).$$

A sample $(Y_1,...,Y_{100})$ has been generated by simulation of ε_i $(i = 1,...,100)$ for $\sigma = 3$. The LSE $\hat{\theta}$

is equal to (1.440,4.346,6.647,-16.329,-74.978)'. The projection of $\hat{\theta}$ on K is $\theta^* = (1.460,1.109,-4.347,7.445,-4.691)', z_{opt} = 45.961$, calculated by the Hildreth-d'Esopo method. The PAS gives the result $\theta_{pas}^* = (1.426,0.879,-3.366,7.598,-6.419), z_{opt}^{(pas)} = 47.927$.
The results are summarized in Fig. 2.
We see that the difference between PAS and the isotonic regression is sufficiently small, at least graphically.

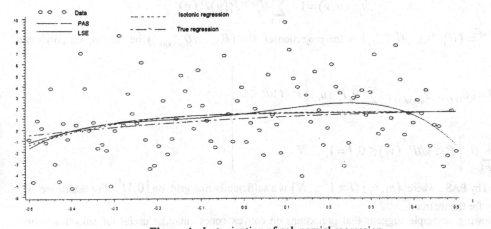

Figure 4 : Isotonization of polynomial regression

5.2 SMOOTHING OF BIVARIATE DENSITY ESTIMATES

Let (ξ, η) be a bivariate random vector distributed according to the density f(s,t) with the marginal distribution functions F_1 and F_2. The transformations $U = F_1(\xi)$ and $V = F_2(\eta)$ have the joint density function $h(u, v)$ on $[0,1]^2$ and its marginal distributions are uniform. $h(u, v)$ is the so-called copula of f(s,t). It represents the dependence structure of (ξ, η) completely and is the base of many nonparametric tests of independence, see [4]. We assume that h is square integrable on $[0,1]^2$ and that it allows a Fourier series representation

$$h(u,v) = 1 + \sum_{i,j=1}^{\infty} \theta_{ij} P_i(u) P_j(v),$$

where $\{P_i; i = 1, ..., \infty\}$ is the system of normed Legendre polynomials on $[0,1]$. A suitable estimator for $h(u,v)$ is (see [4])

$$\hat{h}(u,v) = 1 + \sum_{i,j=1}^{q(n)} \hat{\theta}_{ij} P_i(u) P_j(v), \hat{\theta}_{ij} = n^{-1} \sum_{k=1}^{n} P_i(\frac{R_k}{n+1}) P_j(\frac{S_k}{n+1}),$$

where $(R_k, S_k)(k = 1, ..., n)$ are the bivariate ranks of a sample (ξ_k, η_k) $(k = 1, ..., n)$ and $q(n) = O(n^\alpha), 0 < \alpha < 1$. If ξ and η are positively dependent the inequalities

$$\frac{\partial}{\partial u} \int_0^1 h(u,t)dt \leq 0 \text{ and } \frac{\partial}{\partial v} \int_0^1 h(t,v)dt \leq 0 \quad \forall (u,v) \in [0,1]^2 \tag{3}$$

are true ([4]). The set of all square integrable functions on $[0,1]^2$ satisfying (3) forms a closed convex cone $\tilde{K} \subset L^2[0,1]^2$. The projection h^* of $\hat{h}$ on $\tilde{K}$ is determined by

$$\min_{h \in K} \|\hat{h} - h\|_W^2 = \|\hat{h} - h^*\|_W^2,$$

where $\| \; \|_W$ denotes the norm in the Sobolev space $W_2^{(1)}[0,1]^2$ of functions whose first partial derivatives exist and are in $L^2[0,1]^2$. We approximate h^* by h_{pas}^* in the following way

$$h_{pas}^*(u,v) = 1 + \sum_{i,j=1}^{q(n)} \theta_{ij}^{*(pas)} P_i(u) P_j(v)$$

and $\theta^{*(pas)} = (\theta_{11}^{*(pas)}, ..., \theta_{q(n)q(n)}^{*(pas)})$ is the projection of $\hat{\theta} = (\hat{\theta}_{11}, ..., \hat{\theta}_{q(n)q(n)})$ on the closed polyhedral cone

$$K = \left\{ \begin{array}{l} \theta = (\theta_{11}, ..., \theta_{q(n)q(n)}) : \sum_{i,j=1}^{q(n)} \theta_{ij} P_i^{'}(u_l) \int_0^{v_l} P_j(t)dt \leq 0, \\ \\ \sum_{i,j=1}^{q(n)} \theta_{ij} \int_0^{u_l} P_i(t)dt P_j^{'}(v_l) \leq 0; l = 1, ..., N \end{array} \right\}$$

calculated by PAS , where (u_l, v_l) $(l = 1, ..., N)$ is a sufficiently fine grid on $[0,1]^2$ For details see [4], especially for the matrix C.

The following example suggests that projections on convex cones may be useful for smoothing curve estimates.

Example 2. $(\xi, \eta) \sim N(0, \Sigma), \Sigma = \begin{pmatrix} 1 & 0.2 \\ 0.2 & 1 \end{pmatrix}$.

We got a sample of size 100 by simulation. The grid $\{(u_l, v_l)\}$ contains $m/2 = N = 110$ points. Choosing q=3 we have $p = 9$ The following figures summarize the results and give a report about the progress of PAS.

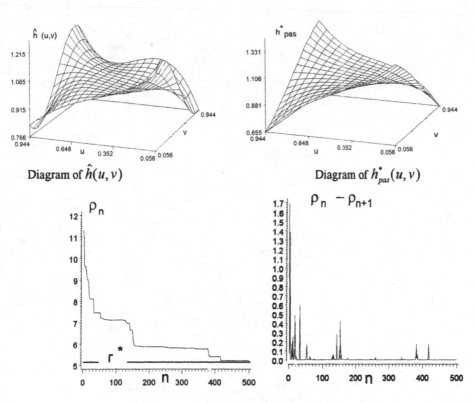

Figure 5: Result and progress of PAS

REFERENCES

1. Barlow, R.E.; Bartholomew, D.J.; Bremner, J.M.; Brunk, H.D. : Statistical Inference under Order Restrictions. Wiley, New York 1972.
2. Fletcher, R. : Practical Methods of Optimization. Wiley , New York 1993
3. Karr, A.F.: Probability.   Springer , New York 1993
4. Rödel, E.: R-Estimation of Normed Bivariate Density Functions. statistics 18(1987), 573-585.
5. Zabinsky, Z.B. ; Smith, R.L.: Pure Adaptive Search in Global Optimization . Mathematical Programming 53(1992), 323-338.

CLASSIFICATION AND DATA MINING

SIMULTANEOUS VISUALIZATION AND CLUSTERING METHODS AS AN ALTERNATIVE TO KOHONEN MAPS

H.H. Bock

RWTH Aachen, Aachen, Germany

ABSTRACT

Kohonen maps are often used for visualizing high-dimensional feature vectors in low-dimensional space. This approach is often recommended for supporting the clustering of data. In this paper an alternative approach is proposed which is more in the lines of multivariate statistics and provides a *simultaneous visualization and clustering* of data. This approach combines projection and embedding methods (such as principal components or multidimensional scaling) with clustering criteria and corresponding optimization algorithms. Four distinct methods are proposed: *projection pursuit clustering* for quantitative data vectors, two *MDS clustering methods* for dissimilarity data (either with or without a representation of classes) and a *group difference scaling method* (known from literature).

1. Introduction

In industry, economics and science we are often faced with the problem of exploring the intrinsic structure of a set of objects (individuals, products, patterns) by analyzing data which describe the properties of those objects, or at least their mutual similarities or dissimilarities. This problem is often approached by *clustering methods*, i.e., by arranging the underlying objects (e.g., the rows of a data matrix) into a suitable number of homogeneous groups or clusters ('types') such that all objects of the same cluster behave similarly with respect to the observed data. A large variety of (partitional or hierarchical) clustering methods has been developed so far for various types of data. They are often based on a pre-specified clustering criterion (thereby producing 'near-optimum' classifications), or may originate from probabilistic models and statistical principles. A full account of relevant literature may be found, e.g., in Bock (1974, 1985, 1996a,b,c), Arabie et al. (1996), or Höppner et al. (1997).

An alternative and very popular approach for getting insight into a data set is provided by *visualization methods*: The objects are represented by points in a (virtual) two- or three- or s-dimensional space such that the (Euclidean) distances between these points reflect approximately the (dis-) similarities which may exist among the investigated objects. A graphical display of these (e.g., two-dimensional) points will then reveal, just by a visual inspection, the structure of the data set, e.g., locate clusters of similar objects, show interesting data trends, or hint to important or outlying cases. Moreover, such a display can be helpful for finding substance-related explanations or interpretations for clusters which were obtained by a purely formal clustering algorithm.

Multivariate data analysis provides a range of classical methods for constructing visual displays:

(a) Methods related to *principal component analysis (PCA)* which assume an $n \times p$ data matrix $(x_{kj}) = (x_1, ..., x_n)'$ where the n feature vectors $x_1, ..., x_n \in R^p$ represent the n objects under study.

(b) Methods related to *multidimensional scaling (MDS)* that start from an $n \times n$ data matrix (d_{kl}) of pairwise dissimilarities d_{kl} among the n objects (and are therefore applicable to the case of qualitative data as well).

Quite recently, there is also a *neural network approach* for tackling with exploratory problems, e.g.:

(c) *Kohonen's self-organizing maps (SOM's)* designed for a quantitative data matrix (x_{kj}) as in (a).

This latter method is described, e.g., in Kohonen (1990, 1991), Varfis & Versino (1992), Ritter et al. (1991), and is used in the clustering framework, e.g., by Murtagh & Hernándesz-Pajares (1995) and Ambroise & Govaert (1996a, b). Thereby, it is often emphasized that Kohonen's SOMs *combine both* the 'clustering' as well the 'visualization' aspects into the same consistent procedure and take into account, in particular, the multidimensional, possibly *non-linear topological structure* of the data under study.

The motivation for this paper is the idea that in order to combine the visualization and clustering aspects of data analysis it is not at all necessary to recur to Kohonen's somewhat indirect and complex procedure which is, e.g., not yet well understood in its topology-retaining properties (see section 2). In contrast, we make the point that several classical and more direct methods from multivariate statistics and cluster analysis might suffice for this task or may be adapted quite easily. So, in this paper, we first comment briefly on Kohonen's SOMs and describe, in the sections 3 to 6, a range of methods related to PCA and MDS that provide a *simultaneous classification and visualization* of data. It seems that these methods provide a convenient alternative to SOM's since they are computationally feasible and based either on precise probabilistic models or on some specific global optimality criterion which facilitates the practical interpretation of the results.

2. Kohonen's self-organizing maps

We set the stage for our data-analytic proposals by briefly describing Kohonen's approach in a geometrical way and will use a terminology that fits more the framework of statistics than that of the common neural network literature:

Suppose that data vectors $x_1, x_2, ..., x_n$ are located in R^p (the input space) near an unknown (possibly virtual) smooth manifold or hypersurface $H \subset R^p$ of a low dimension s, say (the output[1] space). Then the data structure is essentially s-dimensional and can approximately be recovered by the following steps:

a. Find a preliminary (very fine) clustering of the data points $x_1, x_2, ..., x_n$ comprizing a large number m of (typically: small) homogeneous classes $A_1, ..., A_m$ which will be called *mini-clusters* in this article.

b. Since all elements of A_i are concentrated in the same region of R^p, it makes sense to represent each mini-cluster A_i by a suitably chosen point $z_i \in R^p$ that is located near H. (Note that z_i is called a *weight vector* in the neural network terminology.) For example, the centroid $\overline{x}_{A_i} \in R^p$ of all data points from A_i would be appropriate (as it is typical in cluster analysis), but z_i is calculated in another way here (see (2.3), (2.4)).

c. Determine a graph-like 'topological' network (a *map*) in R^p by linking the class representatives $z_1, ..., z_m$ by pairwise edges such that the resulting graph reflects the similarities existing among the mini-clusters, but also recovers the topological connexity among neighbouring mini-clusters inside the unknown (and possibly

bent and winded) manifold H. Typically this map is conceived as a distortion
of an s-dimensional regular rectangular (hexagonal, ...) lattice $\mathcal{L}$ which can be
imagined as some type of coordinate system in the manifold H (Fig. 1).

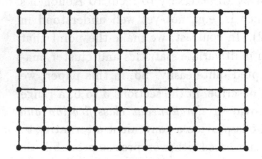

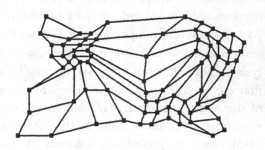

Fig. 1: A 2-dimensional rectangular
lattice $\mathcal{L}$ with $m = 8 \times 10 = 80$ vertices.

Fig. 2: A 2-dimensional Kohonen
map visualizing the m class representatives $z_i \in R^p = R^2$ (or $w_i \in$
$R^s = R^2$).

d. For $p = 2$ a two-dimensional display of the representatives (weight vectors)
$z_1, ..., z_m \in R^p$ together with the suitable links might provide a nice (and ho-
pefully: correct) visualization of the semi-metric and topological structure of the
data (see Fig. 2). For $p > 2$, we must find a suitable low-dimensional representation with dimension $s = 2$ or 3, say, thus m points $y_1, ..., y_m \in R^s$ (the output[2]
space) which approximate as much as possible the geometrical arrangement of
the mini-clusters or the centers $z_1, ..., z_m$ in high-dimensional space R^p, and link
them in the same kind as in $\mathcal{L}$. Typically, a principal component projection of
the points $z_1, ..., z_m$ will be used (Varfis et al. 1992). – Since for two-dimensional
data, the centers z_i, the lattice $\mathcal{L}$ and the projections y_i are all displayed in R^2,
it is often not clear in the literature which one is called the 'output' space; this is
why we introduced 'output[1]' and 'output[2]'.

e. If we dispose, additionally, of information on a secondary *qualitative* property Q
of the individuals (with levels such as color, disease, life-style) such that Q is
(conjectured to be) related to the observed data x_k, we can mark the vertices of
the original lattice $\mathcal{L}$ (or of its distorted versions in R^p or R^s) by corresponding
letters or icons: Then the set of vertices with the same icons may reveal the region
of R^p or R^s where each feature level of Q is concentrated (*semantic map*, Kohonen
1990).

f. When searching for a 'natural' classification of the individuals, the number m
of mini-classes will typically be far too large (see Remark 2.1). Then we will

agglomerate 'similar' (i.e., neighbouring) mini-clusters A_i into some larger (and possibly more relevant) clusters $C_1, ..., C_M$. This latter step seems to be the important one if Kohonen maps are used for clustering purposes, even if it is rarely specified in the literature (but see Murtagh & Hernándes-Pajares 1996).

Remark 2.1: Most applications presuppose a manifold with only $s = 2$ dimensions and use an integer rectangular lattice $\mathcal{L}$ of points $i = (j, l)$ with $j = 1, ..., a$ and $l = 1, ..., b$ (Fig. 1). Then each vertex $i = (j, l) \in \mathcal{L}$ corresponds to one mini-cluster A_i, and the number $m = a \cdot b$ of these mini-clusters will be quite large such that the reduction step f. might be in order.

A Kohonen map (SOM) is constructed by a sequential algorithm that uses the data points in turn: The algorithm starts by assigning to each lattice point $i \in \mathcal{L}$ an initial weight vector $z_i^{(0)} \in R^p$ (typically: a random point in $[0, 1]^p$) and links these vectors by edges in the same way as the vertices i are linked in $\mathcal{L}$. The initial mini-clusters $A_1^{(0)}, ..., A_m^{(0)}$ are all empty. Then the data $x_1, x_2, ...$ are sequentially observed: After observing a new data point x_{n+1} the algorithm updates the previously constructed mini-clusters $A_i^{(n)}$ and their class representatives (weight vectors) $z_i^{(n)} \in R^p$ $(i = 1, ..., m)$, but *it maintains all initially established links* between the centers z_i during the entire process. The updating concerns the center z_{i*} which is closest to x_{n+1} in R^p, but also some other centers which are more or less distant from this latter one *in the graph topology of $\mathcal{L}$*. Closeness in $\mathcal{L}$ is described by a distance index δ on $\mathcal{L} \times \mathcal{L}$ (usually the path distance in $\mathcal{L}$) such that the neigbourhood of some vertex i is defined by $N(i) := \{j \in \mathcal{L} \mid \delta(i, j) \le c\}$ (with some threshold $c \ge 0$). In the following we use, more generally, an 'influence function' $h(i, j)$ which is a decreasing function of the distance $\delta(i, j)$ (with $0 \le h \le 1$). The classical 'crisp' case uses the binary function $h(i, j) = 1$ or 0 if $j \in N(i)$ or not, respectively.

With these specifications, Kohonen's updating algorithm works as follows:

(a) *The new individual $n + 1$ (or x_{n+1}) is assigned to the mini-cluster $A_{i*}^{(n)} \subset \{1, ..., n\}$ that has the closest class representative $z_{i*}^{(n)}$, i.e. with the label:*

$$i^* := \operatorname{argmin}_{i \in \{1, ..., m\}} \{\|x_{n+1} - z_i^{(n)}\|\}. \tag{2.1}$$

(b) *The m class representatives $z_j^{(n)}$ are relocated as follows:*

$$z_{i*}^{(n+1)} := z_i^{(n)} + \alpha_n \cdot (x_{n+1} - z_{i*}^{(n)}) \qquad \text{for } j = i^* \tag{2.2}$$
$$z_j^{(n+1)} := z_j^{(n)} + \alpha_n h(i^*, j)(x_{n+1} - z_j^{(n)}) \qquad \text{for } j \ne i^* \tag{2.3}$$

where the 'learning factor' α_n decreases with n and the factor $h(i^*, j)$ determines the influence of the vertex i^* on the update of the centers corresponding to the neighbouring vertices $j \in \mathcal{L}$.

The lattice $\mathcal{L}$ and its topology enters this algorithm only by the selection of the labels j (neigbours of i^*) where the weight vector $z_j^{(n)}$ is to be modified. It is by far not yet fully clarified which 'topological ordering' in R^p or H may be 'correctly' determined and visualized by this procedure. As far as the clustering point of view is concerned, it appears that:

(a) the minimum-distance assignment of the new data point is similar as in the classical k-means algorithm for minimizing the *variance* or *SSQ clustering criterion*:

$$g_{mn}(\mathcal{C}) := \sum_{i=1}^{m} \sum_{k \in C_i} ||x_k - \bar{x}_{C_i}||^2 \rightarrow \min_{\mathcal{C}}. \qquad (2.4)$$

with respect to all m-partitions $\mathcal{C} = (C_1, ..., C_m)$ of n data points $\{x_1, ..., x_n\}$ from R^p (see, e.g., Bock 1974, §15).

(b) the updating (2.2), (2.3) of the class representatives z_i is analoguous to the approach followed by *stochastic approximation methods* for minimizing the 'continuous' version of the clustering criterion (2.4):

$$\gamma_m(\mathcal{B}, \mathcal{Z}) := \sum_{i=1}^{m} \int_{B_i} ||x - z_i||^2 f(x)dx \rightarrow \min_{\mathcal{B},\mathcal{Z}}. \qquad (2.5)$$

with respect to all m-partitions $\mathcal{B} = (B_1, ..., B_m)$ of the whole input space R^p and all systems $\mathcal{Z} = (z_1, ..., z_m)$ of m class representatives in R^p. Here $f(x)$ is a probability density on R^p and the data points $x_1, x_2, ...$ are assumed to be independent samples from this distribution. This approach has been proposed and investigated as early as in Braverman (1966) and Dorofeyuk (1966) (for details see, e.g., Bock 1974, §29).

(c) the clustering approach is similar to the *sequential version of the k-means algorithm* proposed and investigated by MacQueen (1967) for minimizing (2.5) (see Bock 1974, §29.e).

(d) the underlying criteria (2.4) and (2.5) are related to the well-known *fuzzy clustering criterion*:

$$\kappa(\mathcal{U}, \mathcal{Z}) := \sum_{j=1}^{m} \sum_{k=1}^{n} u_{kj}^{\beta} \cdot ||x_k - z_j||^2 \rightarrow \min_{\mathcal{U},\mathcal{Z}} \qquad (2.6)$$

where the matrix $\mathcal{U} = (u_{ik})_{m \times n}$ denotes a *fuzzy classification* of n individuals such that $u_{ik} \in [0, 1]$ describes the extent (percentage) with which the individual k is assigned to the i-th (fuzzy) class. Here $\beta \geq 1$ is a fixed exponent. For $\beta = 1$ the problem (2.6) reduces to the (non-fuzzy) clustering problem (2.4) (Bock 1974, §15.g). The fuzzy clustering approach is investigated, e.g., in Bock (1979a,b), Bezdek (1981), Höppner et al. (1997); see also Ambroise & Govaert (1996b).

Whilst Kohonen's approach provides very interesting and often useful visualizations of complex structures, it provides no direct clues to the geometrical or interpretational meaning of the resulting (discrete) topology which is, in practice, mostly visualized in the two-dimensional case. Whilst there exist various counter-examples where the 'true' topology is missed, some special type of 'topological correctness' has been proved, e.g., by Ritter & Schulten (1986), Tolat (1990), Jokusch & Ritter (1994) and Fort & Pagès (1996). On the other hand, the method could not yet be motivated by an approach where a global criterion is optimized which incorporates, at the same time, the *clustering* and *visualization* (or *topological*) aspects. In the following sections we will describe some approaches for combining these two latter aspects into one single optimization criterion, and propose corresponding joint clustering and visualization approaches.

3. Projection pursuit clustering

Projection pursuit clustering starts from n high-dimensional quantitative data vectors $x_1, ..., x_n \in R^p$ and constructs a given number m of disjoint (spherical) classes of objects or data points whose centers are concentrated near an optimal low-dimensional hyperplane (or hypersurface) of R^p. Therefore this method reveals classifications which are concentrated along a low-dimensional linear subspace of R^p. If this hyperplane is chosen to be two-dimensional, the results can be visualized in R^2, similarly as with Kohonen maps.

The method is based on the following probabilistic clustering model for n independent random vectors $X_1, ..., X_n$ in R^p:

Hyperplane clustering model:
There exist an unknown *partition* $C = (C_1, ..., C_m)$ of the set $O = \{1, ..., n\}$ of objects with a (fixed known) number $m \geq 2$ of classes, an unknown s-dimensional *hyperplane* $H \in R^p$ (typically $s = 2$ or 3) and a system $Z = (z_1, ..., z_m)$ of unknown *class representatives* $z_1, ..., z_m$ in H (!) such that

$$X_k \sim \mathcal{N}_p(z_i, \sigma^2 \cdot I_p) \qquad \text{for all } k \in C_i \text{ and } i = 1, ..., m. \tag{3.1}$$

Here $\mathcal{N}_p$ denotes the p-dimensional normal distribution, I_p is the $p \times p$ identity matrix and the common variance $\sigma^2 > 0$ may be known or unknown.

The maximum likelihood method for estimating the unknown parameters leads to the following clustering criterion:

$$g(C, H, Z) \quad := \quad \sum_{i=1}^{m} \sum_{k \in C_i} ||x_k - z_i||^2 \rightarrow \min_{\substack{C, H, Z \text{ with } z_i \in H \text{ for all } i}} \tag{3.2}$$

We write $H = a + [v_1, ..., v_s]$ with a vector $a \in R^p$ and an orthonormal system $v_1, ..., v_s, v_{s+1}, ..., v_p$ of vectors in R^p where the first s ones form the matrix $V := (v_1, ..., v_s)$ (with its complement $\overline{V} := (v_{s+1}, ..., v_m)$) and span the linear subspace $[v_1, ..., v_s]$, and denote by $\pi_H(x_k)$ the orthogonal projection of x_k onto H. Then the problem (3.2) may be solved by iterating the following partial minimization steps:

(I) For a fixed pair C and V, we minimize g with respect to $a \in R^p$ and $z_i \in H$, $i = 1, ..., n$. This yields the solutions $a^* = \bar{x} := \frac{1}{n} \sum_{k=1}^{n} x_k$, the centroid of all data points, and $z_i^* = \pi_H(\bar{x}_{C_i})$, i.e., the projection of the centroid $\bar{x}_{C_i}$ of C_i, i.e. of all x_k with $k \in C_i$. It remains to minimize the criterion:

$$g(C, V) := \sum_{i=1}^{m} |C_i| \cdot \|\bar{x}_{C_i} - \pi_H(\bar{x}_{C_i})\|^2 + \sum_{i=1}^{m} \sum_{k \in C_i} \|x_k - \bar{x}_{C_i}\|^2$$

$$= tr(\overline{V}' B(C)\overline{V}) + tr(W(C)) \tag{3.3}$$

$$= tr(V' S V) + tr(V'W(C)V) \tag{3.4}$$

$$\underset{C,V}{\longrightarrow} \min$$

where we recall the decomposition formula for the $p \times p$ scatter matrices $W(C), B(C)$ *in* and *between* the classes of C, respectively:

$$S := \sum_{k=1}^{n} (x_k - \bar{x})'(x_k - \bar{x}) = W(C) + B(C). \tag{3.5}$$

(II) For a fixed partition C, we optimize (3.3) with respect to the matrix V or, equivalently, to the complementary matrix $\overline{V}$, i.e. we solve $tr(\overline{V}' B(C)\overline{V}) \to \min \overline{V}$. It appears that this latter problem corresponds to a weighted principal component analysis for the m vectors $\bar{x}_{C_1} - \bar{x}, ..., \bar{x}_{C_m} - \bar{x} \in R^p$ and results in the solution $H^* = \bar{x} + [v_1, ..., v_s]$ where $v_1, ..., v_s$ are the eigenvectors of $B(C)$ belonging to the s *largest* eigenvalues of that matrix. Thus H is a *modified principal component hyperplane*.

(III) Finally, for a fixed hyperplane H (i.e., for a fixed V) we optimize (3.4) with respect to the m-partition C. Since the first term in (3.4) is constant here, we have to minimize the second term $tr(V'W(C)V$, this is exactly the clustering criterion (2.4) (variance criterion) for the *projected* data points $\pi_H(x_1), ..., \pi_H(x_n)$. This clustering problem can be approximately solved, e.g., by the 'k-means algorithm' (Bock 1974).

Taking all this together, we obtain the following *projection pursuit clustering algorithm* for simultaneously determining an optimum hyperplane and an optimal classification of the data (Bock 1987a): We start with an arbitrary initial m-partition $C^{(0)}$ of OIt and iterate cyclically the partial minimization steps (I), (II), (III). Obviuosly, this results in a sequence of partitions $C^{(t)}$, hyperplanes $H^{(t)}$ and center systems $Z^{(t)}$ $(t = 0, 1, ...)$ with steadily decreasing values for $g(C, H, Z)$ (even if in (III) only a sub-optimum partition might be used). The algorithm attains its stationary state typically after some few iterations.

Remark 3.1: It is obvious that, and how, this method can be generalized in order to comply with other cluster types (e.g., of ellipsoidal instead of a spherical shape) or with s-dimensional *non-linear hypersurfaces* instead of hyperplanes (see also the paper of Ritter (1997) in this volume). Bock (1987a) describes an extension where *common* 'factors' or 'dimensions' are shared by all classes.

4. MDS-based clustering and visualization

Projection pursuit clustering works for quantitative data since the clustering criterion (3.2) makes sense in this case. However, (3.2) does not apply for other types of data which, nevertheless, are often met in practice such as qualitative or symbolic data, or a mixture of various data types. In order to tackle with these more general situations, we propose an approach based on *multidimensional scaling* (MDS): Thus, in a first step, we calculate, for each pair of objects $k, l \in \mathcal{O}$, a real-valued measure d_{kl} which characterizes the dissimilarity between these objects. Thereby we obtain a data matrix $D = (d_{kl})_{n \times n}$ with the usual properties $0 = d_{kk} \leq d_{kl} = d_{lk} < \infty$ for all k, l.

Based on this dissimilarity matrix D, we construct, in a second step, *simultaneously*,

- an 'optimal' m-partition $\mathcal{C} = (C_1, ..., C_m)$ of the n objects and
- a configuration of points $x_1, ..., x_n$ in a Euclidean space of dimension s, say, such that their Euclidean distances $\delta_{kl} := \|x_k - x_l\|$ approximate optimally the given dissimilarities d_{kl}.

We begin by recalling some basic facts from (MDS).

4.1 Classical multidimensional scaling methods

Given a matrix $D = (d_{kl})$ of dissimilarities, MDS looks for n points $x_1, ..., x_n \in R^s$, i.e., a configuration $X = (x_1, ..., x_n)' \in R^{n \times s}$ such that D is 'best' approximated by $\Delta(X) := (\delta_{kl}) = (\|x_k - x_l\|)$. Insofar the configuration X will provide an s-dimensional display of the mutual similarities and dissimilarities existing among the n objects which is easily visualized for $s = 2$ or $s = 3$.

The formal definition of this problem specifies first a goodness-of-fit criterion $g(X, D)$ and then minimizes this criterion with respect to the configuration X (where we may assume $\bar{x} = 0$ without restriction of generality). Two well-known criteria are given by:

$$STRESS: \quad g_1(X, D) \quad := \quad \sum_{k=1}^{n} \sum_{l<k} (d_{kl} - \|x_k - x_l\|)^2 \to \min_X \qquad (4.1)$$

$$SSTRESS: \quad g_2(X, D) \quad := \quad \sum_{k=1}^{n} \sum_{l<k} (d_{kl}^2 - \|x_k - x_l\|^2)^2 \to \min_X. \qquad (4.2)$$

They are minimized by suitable numerical methods such as steepest gradient, majorization methods, tunneling methods etc. (see, e.g., Mathar 1994, Groenen & Heiser 1996, Borg & Groenen 1997). D is called *Euclidean* if there exist, for some dimension $s \geq 1$, n points $x_1, ..., x_n \in R^s$ with $d_{kl} = \|x_k - x_l\|$ for all k, l (i.e. if the attainable

minimum error in (4.1) or (4.2) reduces to zero). A famous theorem of Menger states
that D is Euclidean if and only if the matrix $S = (s_{kl})_{n \times n}$ with entries

$$s_{kl} := -\tfrac{1}{2}(d^2_{kl} - \overline{d^2_{k,\cdot}} - \overline{d^2_{l,\cdot}} + \overline{d^2_{\cdot,\cdot}}) \tag{4.3}$$

is positive definite (then the minimum dimension s is given by the rank of S). Here the
bar denotes averaging the squares d^2_{kl} over the index, or indices, replaced by a dot.

It appears that in the Euclidean case we have $s_{kl} = (x_k - \bar{x})'(x_l - \bar{x})$ for all k, l such
that we will denote the numbers s_{kl} *pseudo-scalars* here. Even in the general case, the
equations (4.3) allow a simple inversion formula:

$$d^2_{kl} = s_{kk} + s_{ll} - 2s_{kl} \tag{4.4}$$

such that the information contained in $D = (d_{kl})$ and $S = (s_{kl})$ is equivalent.

This fact leads to the idea to approximate, as an alternative to (4.1) and (4.2), the
pseudo-scalars s_{kl}, (4.3), by the scalar products $(x_k - \bar{x})'(x_l - \bar{x}) = x'_k x_l$ of a suitable
configuration X (where we recall the irrelevant constraint $\bar{x} = 0$). This leads to the
scalar version of MDS specified by:

$$STRAIN: \quad g_3(X, S) := \sum_{k=1}^{n} \sum_{l<k} (s_{kl} - x'_k x_l)^2 = tr([S - XX']^2) \to \min_{X}. \tag{4.5}$$

An analytical solution of this problem has been given by Keller (1962) and Mathar
(1985; for a generalized, orthogonally invariant matrix norm instead of the trace). It is
provided by the spectral decomposition $S = \sum_{t=1}^{n} \lambda_t v_t v'_t$ of the matrix S where its *non-
zero* eigenvalues are labeled in decreasing order: $\lambda_1 \geq \cdots \geq \lambda_r > 0 > \lambda_{r+1} \geq \cdots \geq \lambda_{r+q}$
(thus r and q are the number of positive and negative eigenvalues of S, respectively):

Theorem 4.1: *An optimum configuration is given by:*

$$X^* = (x_1^*, ..., x_n^*)' \quad = \quad (\sqrt{\lambda_1} v_1, ..., \sqrt{\lambda_s} v_s) \tag{4.6}$$

if $s \leq r$, and by:

$$X^* = (x_1^*, ..., x_n^*)' \quad = \quad (\sqrt{\lambda_1} v_1, ..., \sqrt{\lambda_r} v_r, 0, ..., 0) \tag{4.7}$$

in the case $s > r$ (where $s - r$ zero columns are to be introduced).

4.2 A combined clustering and visualization strategy

A method for simultaneously clustering and visualizing the n objects on the basis
of the observed dissimilarities d_{kl} may be obtained by combining the MDS criterion
(4.5) with the variance criterion (2.4) of cluster analysis: We look for an m-partition

$C = (C_1, ..., C_m)$ of $\mathcal{O}$ and a configuration $X = (x_1, ..., x_n)'$ of points in R^s such that the criterion

$$g(C, X) \quad := \quad \sum_{k=1}^{n}\sum_{l=1}^{n}(s_{kl} - x_k'x_l)^2 + 2\beta \cdot \sum_{i=1}^{m}\sum_{k \in C_i} ||x_k - \bar{x}_{C_i}||^2 \qquad (4.8)$$

is minimized with respect to C and X (Bock 1987a). Here $\beta > 0$ is a weight parameter that controls the trade-off between the aspects of visualization (Euclidean approximation) and clustering and can be suitably specified. – A little algebra shows that

$$g(C, X) \quad := \quad tr(S^2 - S_C) + tr([S_C - XX']^2) \qquad (4.9)$$

with the modified S-matrix:

$$S_C \quad := \quad S + \beta \cdot (R_C - I_n) \leq S \qquad (4.10)$$

where

$$R_C \quad := \quad \left(\begin{array}{ll} r_{kl} = 1/|C_i| & \text{for } k, l \in C_i \\ r_{kl} = 0 & \text{else} \end{array} \right) \in R^{n \times n}. \qquad (4.11)$$

The two representations (4.8) and (4.9) for g show that an iterative alternating minimization device is possible:

(I) For a given partition C, we minimize (4.9) with respect to X by solving $tr([S_C - XX']^2) \to \min_X$. The solution is given by Theorem 4.1 that shows that the optimum configuration X^* is found by the spectral decomposition of the modifies S-matrix S_C. Thus step (I) corresponds to a *modified scalar version of MDS*.

(II) For a given configuration X, we optimize (4.8) with respect to the partition C: This amounts to minimizing the variance criterion $\sum_{i=1}^{m}\sum_{k \in C_i} ||x_k - \bar{x}_{C_i}||^2 = tr(W(C))$, a problem that can be (approximately) solved by the k-means algorithm (see section 2 or Bock 1974) to the given configuration $x_1, ..., x_n \in R^s$.

From (I) and (II) we obtain the following *combined MDS clustering method*: We start with an arbitrary initial partition $C^{(0)}$ and iterate cyclically the minimization steps (I) and (II). This process generates a series of configurations $X^{(t)}$ and partitions $C^{(t)}$ ($t = 0, 1, ...$) with steadily decreasing criterion values $g(C, X)$ (see Bock 1986, 1987b). After having attained a stationary state, the algorithm yields a simultaneous clustering and visualization of the underlying n objects which can be displayed graphically as in Kohonen's method, but for a quite general type of data.

Remark 4.1: In (4.8) we have used the STRAIN criterion g_3 from (4.5). Similar MDS clustering methods result if we use instead the STRESS g_1 or SSTRESS g_2 criteria and apply numerical minimization algorithms in the step (I) instead of the explicit solution provided by Theorem 4.1.

Remark 4.2: The choice of a suitable weight β might be controversal in practice and several values for β must be tried. As an alternative, we can use a cross-validation method or select the weight which minimizes (4.8) with respect to β as well.

5. Combining clustering with the simultaneous visualization of objects and classes

When classifying objects on the basis of a dissimilarity matrix $D = (d_{kl})$ it may be useful not only to display the n *objects* in R^s (usually in R^2 or R^3), but also to find suitable representative points for the constructed *classes* $C_1, ..., C_m$ in this same space R^s. Thus, in this section, we look

- for a suitable m-partition $\mathcal{C} = (C_1, ..., C_m)$ of the objects,
- for n points $x_1, ..., x_n \in R^s$ which represent these n objects
- and, simultaneously, m points $y_1, ..., y_m \in R^s$ which represent the m classes of $\mathcal{C}$.

A first (and often used) attempt proceeds by (1) calculating an optimum partition for the data, moreover and *independently from (1)*, by (2) determining an optimum low-dimensional representation $x_1^*, ..., x_n^* \in R^s$ for the objects by MDS methods (section 4.1) and then (3) defining class representatives as the centroids $y_i := \overline{x}_{C_i}^* \in R^s$ of the classes C_i of the given partition $\mathcal{C}$. This approach neither incorporates explicitly the relationship which may exist between the classes as expressed, e.g., by average dissimilarities $\overline{D}_{C_i, C_j}$ between the classes nor takes it into account the existence of clusters when looking for the configuration X^*.

In the following subsections 5.1 and 5.2 we consider a *simultaneous* visualization method for objects *and* classes and incorporate it into the clustering framework in 5.3.

5.1 Average dissimilarities between classes, adjusted average dissimilarities

Consider a set of n objects together with a dissimilarity matrix $(d_{kl})_{n \times n}$ and an m-partition $\mathcal{C} = (C_1, ..., C_m)$ of $\mathcal{O}$. For characterizing the dissimilarity between classes we may define the following concepts (Bock 1986, 1987a,b):

- *The average (squared) class dissimilarity between C_i and C_j:*

$$\overline{D_{C_i,C_j}^2} \quad := \quad \frac{1}{|C_i| \cdot |C_j|} \sum_{k \in C_i} \sum_{l \in C_j} d_{kl}^2 \tag{5.1}$$

which is, for $i = j$, to be interpreted as a heterogeneity measure for C_i. We will use the notation

$$\overline{D_{C_i,\cdot}^2} \quad := \quad \frac{1}{n \cdot |C_i|} \sum_{k \in C_i} \cdot \sum_{l=1}^{n} d_{kl}^2 = \sum_{j=1}^{m} \frac{|C_i|}{n} \cdot \overline{D_{C_i,C_j}^2} \tag{5.2}$$

for the weighted average of these average dissimilarities. Similarly $\overline{D_{\cdot,\cdot}^2} := \overline{d_{\cdot,\cdot}^2} = \sum_k \sum_l d_{kl}^2 / n^2$.

- *The adjusted average (squared) class dissimilaritiy between C_i and C_j:*

$$\delta_{C_i,C_j} \quad := \quad \overline{D^2_{C_i,C_j}} - \frac{1}{2}\overline{D^2_{C_i,C_i}} - \frac{1}{2}\overline{D^2_{C_j,C_j}} \tag{5.3}$$

with $\delta_{C_i,C_i} = 0$ and analoguously weighted averages such as $\overline{\delta_{C_i,\cdot}}$ and $\overline{\delta_{\cdot,\cdot}}$.

- *The aggregated pseudo-scalars*

$$\overline{S_{C_i,C_j}} \quad := \quad \frac{1}{|C_i| \cdot |C_j|} \sum_{k \in C_i} \sum_{l \in C_j} s_{kl}. \tag{5.4}$$

It appears that these latter values can be obtained by a weighted centering of the rows and colums of the $m \times m$ matrices $(\overline{D^2_{C_i,C_j}})$ and (δ_{C_i,C_j}) according to:

$$\overline{S_{C_i,C_j}} \quad = \quad -\frac{1}{2}(\overline{D^2_{C_i,C_j}} - \overline{D^2_{C_i,\cdot}} - \overline{D^2_{C_j,\cdot}} + \overline{D^2_{\cdot,\cdot}}) \tag{5.5}$$

$$\overline{S_{C_i,C_j}} \quad = \quad -\frac{1}{2}(\delta_{C_i,C_j} - \overline{\delta_{C_i,\cdot}} - \overline{\delta_{C_j,\cdot}} + \overline{\delta_{\cdot,\cdot}}). \tag{5.6}$$

In analogy to (4.4) there exists an inversion formula for (5.6):

$$\delta_{C_i,C_j} \quad = \quad \overline{S_{C_i,C_i}} + \overline{S_{C_j,C_j}} - 2 \cdot \overline{S_{C_i,C_j}}, \tag{5.7}$$

but *not* for (5.5).

Remark: The interpretation of these concepts is obvious if we consider the case of a Euclidean dissimilarity matrix (d_{kl}) with an (exact!) representation by the configuration $X = (x_1, ..., x_n)'$ (see section 4.1). Then

$$\frac{1}{2}\overline{D^2_{C_i,C_i}} \quad = \quad var(C_i) := \frac{1}{n_i}\sum_{k \in C_i} ||x_k - \bar{x}_{C_i}||^2 \tag{5.8}$$

is the empirical variance in the class C_i,

$$\overline{D^2_{C_i,C_j}} \quad = \quad var(C_i) + var(C_j) + ||\bar{x}_{C_i} - \bar{x}_{C_j}||^2, \tag{5.9}$$

whereas

$$\delta_{C_i,C_j} \quad = \quad ||\bar{x}_{C_i} - \bar{x}_{C_j}||^2 \tag{5.10}$$

is the squared Euclidean distance between the class centroids. This latter fact shows that it might be misleading to base any MDS considerations involving classes on the average dissimilarities $\overline{D^2_{C_i,C_j}}$ if the Euclidean metric is used in R^s. Instead they should be based on the adjusted dissimilarities δ_{C_i,C_j} or, equivalently, on the aggregated pseudo-scalars $\overline{S_{C_i,C_j}}$.

5.2 Simultaneous MDS for objects and classes

The last remark shows that, when starting with a dissimilarity matrix (d_{kl}) and a fixed m-partition $\mathcal{C}$, a *simultaneous* s-dimensional representation $x_1, ..., x_n, y_1, ..., y_m \in R^s$ of objects and classes can be tried either

(1) by approximating the dissimilarities d_{kl}^2, δ_{C_i,C_j} and $\delta_{C_i,\{k\}}$ optimally by the corresponding Euclidean distances $||x_k - x_l||^2$, $||\bar{x}_{C_i} - \bar{x}_{C_j}||^2$ and $||\bar{x}_{C_i} - x_k||^2$, respectively, or

(2) by approximating the pseudo-scalars s_{kl}, $\overline{S_{C_i,C_j}}$ and $\overline{S_{C_i,\{k\}}}$ optimally by the corresponding scalar products $x_k' x_l$, $y_i' y_j$ and $y_i' x_k$, respectively.

We will follow this latter approach here and consider the *goodness-of-fit criterion* (Bock 1986, 1987a,b):

$$g_4(X,Y,\mathcal{C}) \quad := \quad \sum_{k=1}^{n} \sum_{l=1}^{n} (s_{kl} - x_k' x_l)^2 \tag{5.11}$$

$$+ 2\alpha^2 \sum_{i=1}^{m} \sum_{k=1}^{n} (\overline{S_{C_i,k}} - y_i' x_k)^2 + \alpha^4 \sum_{i=1}^{m} \sum_{j=1}^{m} (\overline{S_{C_i,C_j}} - y_i' y_j)^2 \rightarrow \min_{X,Y}.$$

which is to be minimized with respect to the configurations X, Y (for a fixed m-partition $\mathcal{C}$. Some elementary matrix calculations show that this criterion is identical to the STRAIN criterion (4.5) where X and S have to be replaced by the $(n+m) \times p$ matrix $\tilde{X} := (X \; Y)'$ and the $(n+m) \times (n+m)$ matrix

$$\tilde{S}(\mathcal{C}) := \left(\begin{array}{cc} s_{kl} & \overline{S_{\{k\},C_j}} \\ \hline \overline{S_{C_i,\{l\}}} & \overline{S_{C_i,C_j}} \end{array} \right),$$

respectively. An analytical solution of the minimization problem (5.11) is provided by:

Theorem 5.1:

(a) *Consider the STRAIN version (4.5) of the MDS problem for the symmetric $n \times n$ matrix*

$$\hat{S} \quad := \quad (I_n + \alpha^2 M_C M_C')^{1/2} \cdot S \cdot (I_n + \alpha^2 M_C M_C')^{1/2}$$

where $M_C = (M_{ij})_{m \times m}$ is the $n \times n$ matrix with column vectors $M_{ii} = (1, 1, ..., 1)'/|C_i| \in R^{|C_i|}$ and $M_{ij} = 0 \in R^{|C_i|}$ for $i \neq j$. Its solution is given by the configuration

$$\hat{X} = (\hat{x}_1, ..., \hat{x}_n)' \quad = \quad (\sqrt{\lambda_1} v_1, ..., \sqrt{\lambda_s} v_s) \tag{5.12}$$

where λ_i and v_i are the eigenvalues and eigenvectors of $\hat{S}$, respectively (assuming $s \leq \hat{r}$, the number of positive eigenvalues of $\hat{S}$).

(b) *From this matrix, the optimum configuration for the visualization problem (5.11) is given by*

$$X^* = (x_1^*, ..., x_n^*)' \; := \; (I_n + \alpha^2 M_C M_C')^{-1/2} \hat{X} \tag{5.13}$$

$$Y^* = (y_1^*, ..., y_m^*)' \; = \; M_C' X^* = (\bar{x}_{C_1^*}, ..., \bar{x}_{C_m^*}). \tag{5.14}$$

Thus, for the criterion (5.11), the optimum class representatives y_i^* happen just to be the class centroids of the resulting configuration $x_1^*, ..., x_n^*$. Note, however, that in contrast to the approach (1), (2), (3) sketched at the beginning of this section 5, these latter points are determined here in dependence on the given partition C.

5.3 A combined MDS clustering and visualization method

It is obvious that the previous theorem provides a *combined clustering and visualization method* if we combine, similarly as in (4.8), the criterion $g_4(X, Y, C)$, (5.11), linearly with a clustering criterion based on dissimilarities such as, e.g., $g_5(C) := \sum_{i=1}^{m} \sum_{j=1}^{m} \overline{S}_{C_i, C_j}$, (with a suitable weight $\beta > 0$) and minimize the resulting criterion $\tilde{g} := g_4 + \beta g_5$ with respect to X, Y *and* the partition C as well. Again an approximate solution can be found by applying an alternating minimization algorithm where, in a first step, *optimum configurations X and Y* are obtained (from Theorem 5.1) and in a second step an *optimum clustering* is found (or approximated) by applying an exchange algorithm to the criterion $\tilde{g}$ (for fixed configurations X and Y).

6. Group difference scaling

There are several other attempts in the framework of MDS in order to formulate combined clustering and visualization criteria. For example, Heiser (1993) considers average distance of the form

$$\overline{D}_{C_i, C_j} \; := \; \frac{1}{|C_i| \cdot |C_j|} \sum_{k \in C_i} \sum_{l \in C_j} d_{kl} \tag{6.1}$$

and looks for a partition $C = (C_1, ..., C_m)$ of $\mathcal{O}$ and m class representatives $z_1, ..., z_m \in R^s$ such that the criterion

$$g(Y, C) \; := \; \sum_{1 \le i < j \le m} \sum_{k \in C_i} \sum_{l \in C_j} (d_{kl} - ||y_i - y_j||)^2 \tag{6.2}$$

$$:= \; \sum_{i<j} |C_i||C_j| (\overline{D}_{C_i, C_j} - ||y_i - y_j||)^2 + \sum_{i<j} \sum_{k \in C_i} \sum_{l \in C_j} (d_{kl} - \overline{D}_{C_i, C_j})^2 \tag{6.3}$$

$$=: \; g_6(C, Y) + g_7(C) \; \to \; \min_{C, Y} .$$

This criterion formulalizes the idea that, for each pair of classes C_i, C_j, all pairwise dissimilarities d_{kl} between objects $k \in C_i$, $l \in C_j$ are (or should be) about the same: $d_{kl} \approx \tau_{ij}$, and this common class-specific value τ_{ij} is given by the Euclidean distance

$\tau_{ij} = \|y_i - y_j\|$ of the class representatives $y_i, y_j \in R^s$. The decomposition (6.3) shows that this criterion combines a term $g_6(\mathcal{C}, Y)$ quantifying the visualization (goodness-of-fit) aspect for the classes with a term $g_7(\mathcal{C})$ measuring the heterogeneity of the classes in terms of dissimilarities d_{kl} (this is different from (4.8) where we have used the corresponding Euclidean points $x_k \in R^s$).

A corresponding alternating minimization algorithm iterates the following two steps:

(I) For a fixed partition $\mathcal{C}$, we minimize the criterion $g_6(\mathcal{C}, Y)$ with respect to the class representatives $y_1, ..., y_m \in R^s$. This is a weighted MDS problem of the type (4.1) which can be solved by standard MDS algorithms (see Heiser 1993, Heiser & Groenen 1997).

(II) For a fixed system of class representatives Y, we minimize $g(\mathcal{C}, Y)$ with respect to the partition $\mathcal{C}$ by using a pairwise exchange algorithm: Starting with an initial partition $\mathcal{C}^{(0)}$, we successively transfer objects $s \in C_r$ to another class C_t provided that this exchange reduces the criterion $g(\mathcal{C}, Y)$. It can be shown that this is the case if

$$\Delta := \tau_{rt}^2 + \sum_{\substack{i=1 \\ i \neq r}}^{m} \sum_{k \in C_i} (d_{ks} - \tau_{ir})^2 - \sum_{\substack{i=1 \\ i \neq t}}^{m} \sum_{k \in C_i} (d_{ks} - \tau_{it})^2 > 0. \tag{6.4}$$

Obviously, the criterion (6.2) emphasizes the visual representation of the *classes* only and neglects the representation of the n *objects*. For visualizing these latter ones as well, Heiser proposes an 'add-a-point' technique where for each object $k \in \mathcal{O}$ a representative point $x_k \in R^s$ is constructed that minimizes a criterion of the type $\sum_{i=1}^{m} \sum_{l \in C_i} (d_{kl} - \|x_k - y_i\|)^2$.

7. Conclusions

In this paper we have shown that classical clustering and scaling methods can be combined in order to attain a simultaneous visualization and classification of data. In contrast to Kohonen's maps, these methods are based on a well-specified goodness-of-fit criterion that is iteratively minimized by alternatingly projecting and clustering points in some Euclidean space. Some related approaches are described in Ultsch (1993), Krzanowski (1994) and Ambroise & Govaert (1996a,b).

References:

Ambroise, Ch., G. Govaert (1996a): Constrained clustering and Kohonen self-organizing maps. J. of Classification 13, 299-313.

Ambroise, Ch., G. Govaert (1996b): Analysing dissimilarity matrices via Kohonen maps. Lecture given at the 5th Conference of the International Federation of Classification Societies, Kobe/Japan, March 1996, Abstract volume II, p. 96-99.

Arabie, Ph., Hubert, L. and G. De Soete (eds.) (1996): Clustering and classification.

World Science Publishers, River Edge/NJ.

Bezdek, J.C. (1981): Pattern recognition with fuzzy objective function algorithms. Plenum Press, New York.

Bock, H.H. (1974): Automatische Klassifikation (Cluster-Analyse). Vandenhoeck & Ruprecht, Göttingen.

Bock, H.H. (1979a): Clusteranalyse mit unscharfen Partitionen. In: Bock, H.H. (ed.): Klassifikation und Erkenntnis III: Numerische Klassifikation. Indeks-Verlag, Frankfurt, 137-163.

Bock, H.H. (1979b): Fuzzy clustering procedures. In: R. Tomassone (ed.): Analyse des donées et informatique. Institut de Recherche en Informatique et en Automatique (INRIA), Le Chesnay, France, 205-218.

Bock, H.H. (1985): On some significance tests in cluster analysis. J. of Classification 2, 77-108.

Bock, H.H. (1986): Multidimensional scaling in the framework of cluster analysis. In: P.O. Degens, H.-J. Hermes, O. Opitz (Eds.): Classification and its environment. Studien zur Klassifikation no. 17. Indeks-Verlag, Frankfurt, 1986, 247-258.

Bock, H.H. (1987a): On the interface between cluster analysis, principal component analysis, and multidimensional scaling. In: H. Bozdogan, A.K. Gupta (Eds.): Multivariate statistical modeling and data analysis. D. Reidel, Dordrecht, 1987, 17-34.

Bock, H.H. (1987b): Metrische Modelle bei der Klassifikation mit Unähnlichkeitsmatrizen. In: H. Iserman et al. (Eds.): Operations Research Proceedings 1986. Springer-Verlag, Berlin, 1987, 440-446.

Bock, H.H. (1996a): Probabilistic models in partitional cluster analysis. In: A. Ferligoj and A. Kramberger (eds.): Developments in data analysis. FDV, Metodoloski zvezki, 12, Ljubljana, Slovenia, 1996, 3-25.

Bock, H.H. (1996b): Probabilistic methods in cluster analysis. Computational Statistics and Data Analysis 3, 5-28.

Bock, H.H. (1996c): Probability models and hypotheses testing in partitioning cluster analysis. In: P. Arabie, L. Hubert and G. De Soete (eds.): Clustering and classification. World Science Publishers, River Edge/NJ, 1996, 377-453.

Bock, H.H. (1997): Simultaneous clustering and visualization methods with a view towards Kohonen's neural networks. In: G. Della Riccia (ed.): Learning, networks and statistics. Proc. ISSEK Workshop, University of Udine, September 1996. CISM Courses and Lectures, Springer-Verlag, Wien, 1997 (in press).

Borg, I., Groenen, P. (1997): Modern multidimensional scaling: Theory and applications. Springer-Verlag, New York.

Braverman, E.M. (1966): The method of potential functions in the problem of training machines to recognize patterns witout a teacher. Automation Remote Control 27, 1748-1771.

Dorofeyuk, A.A. (1966): Teaching algorithms for a pattern recognition machine witout a teacher based on the method of potential functions. Automation Remote Control 27, 1728-1737.

Fort, J.-C., Pagès, G. (1996): About the Kohonen algorithm: Strong or weak selorganization? Neural networks 9, 773-785.

Groenen, P.J.F., Heiser, W.J. (1996): The tunneling method for global optimization in multidimensional scaling. Psychometrika 61, 529-550.

Heiser, W. (1993): Clustering in low-dimensional spaces. In: O. Opitz, B. Lausen, R. Klar (Eds.): Information and classification - Concepts, methods and applications. Springer-Verlag, Heidelberg, 1993, 162-173.

Heiser, W., Groenen, P. (1997): Cluster differences scaling with a within-clusters loss component and a fuzzy successive approximation strategy to avoid local minima. Psychometrika 62, 63-83.

Höppner, F., Klawonn, F. and R. Kruse (1997): Fuzzy-Clusteranalyse. Verfahren für die Bilderkennung, Klassifizierung und Datenanalyse. Verlag Vieweg, Wiesbaden.

Jokusch, S., Ritter, H. (1994): Self-organizing maps: Local competition and evolutionary optimization. Neural Networks 7, 1229-1239.

Keller, J.B. (1962): Factorization of matrices by least squares. Biometrika 49, 239-242.

Kohonen, T. (1990): The self-organizing map. Proceedings of the IEEE 78, 1464 - 1480.

Kohonen, T. (1991): Artificial neural networks 1,2. North Holland, Amsterdam.

Krzanowski, W.J. (1994): Ordination in the presence of group structure for general multivariate data. J. of Classification 11, 195-207.

MacQueen, J. (1967): Some methods for classification and analysis of multivariate observations. In: L. LeCam, J. Neyman (eds.): Proc. 5th Berkeley Symp. Math, Statist. Prob. 1965/66. Univ. California Press, Berkeley, 1967, Vol. 1, 281-297.

Mathar, R. (1985): The best Euclidean fit to a given distance matrix in prescribed dimensions. Linear Algebra and its Applications 67, 1-6.

Mathar, R. (1994): Multidimensional scaling with l_p-distances, a unifying approach. In: H.H. Bock, W. Lenski, M.M. Richter (eds.): Information sytsems and data analysis. Springer-Verlag, Heidelberg, 325-331.

Murtagh, F., Hernández-Pajares, M. (1995): The Kohonen self-organizing map method: an assessment. J. of Classification 12, 165-190.

Ritter, H. (1997): Neural networks for rapid learning in computer vision and robotics. This volume.

Ritter, H., Schulten, K. (1986): On the stationary state of Kohonen's self-organizing sensory mapping. Biological Cybernetics 54, 99-106.

Ritter, H., T. Martinetz and K. Schulten (1991): Neuronale Netze. Eine Einführung in die Neuroinformatik selbstorganisierender Netzwerke. Addison-Wesley, Bonn.

Tolat, V.V. (1990): An analysis of Kohonen's self-organizing maps using a system of energy functions. Biological Cybernetics 64, 155-164.

Ultsch, A. (1993): Self-organizing neural networks for visualisation and classification. In: O. Opitz, B. Lausen, R. Klar (Eds.): Information and classification - Concepts, methods and applications. Springer-Verlag, Heidelberg, 1993, 301-306.

Varfis, A., Versino, C. (1992): Clustering of socio-economic data with Kohonen maps. International Journal on Neural and Mass-Parallel Computing and Information Systems "Neural Network World", 2, 813-833.

DATA ANALYSIS IN INDUSTRY
A PRACTICAL GUIDELINE

H. Hellendoorn
Siemens AG, Munich, Germany

Abstract

We present a methodology for data mining in large industrial data sets. We show the most important steps in the process of extracting information out of the data: (1) Preprocessing the data, (2) Reducing the data, (3) Modeling the data, (4) Rule extraction, and (5) Drift analysis. We also show some examples of industrial applications where fuzzy logic based data analysis plays a role.

1   Introduction

Fuzzy logic [4] is often being used in connection with sensor analysis, as experiences
in chemical industry, paper industry, car industry, traffic control, etc. show. There
are several reasons for the increasing interest in this area. The main reasons are the
increased use of sensors in any industrial area and the wish to describe and understand
processes in an easy rule-based way. In particular in areas where many input parameters play a role, fuzzy logic can help to obtain more transparency in the description of
the processes. Another argument for the use of fuzzy logic in sensor analysis is the im-
precision that is allowed: perhaps a purely mathematical approach with many formulas
might do better, but to handle such a theory often demands too many assumptions,
that weaken this approach severely.

In this short paper we will briefly describe the general method of data analysis, and
show with the help of some examples the use of it.

2   The Methodology

Although each project in which data analysis plays a role is different, there are some
similarities in the development steps of these projects. Of course, there are differences
too, time-dependent data can be treated in a completely different way than time-
independent data. Usually we follow the steps presented below in data mining projects
[1].

We will describe a five step approach to data analysis, viz.

1. Preprocessing the data, also called data reconciliation: The data comes from
   sensors that do not always function well. Hence the data should be 'cleaned'.

2. Reducing the data. Usually there are many sensors generating megabytes of data.
   The system should reduce the number of inputs and the amount of data.

3. Modeling the data: There are several ways to model the data. This model can
   be used forecast and control process behavior.

4. Rule extraction: The behavior of the process is described in terms of causal rules
   and presented to the operator in an understandable form.

5. Drift analysis: It is particularly interesting for the operator to know in which
   state the machine is situated.

It is important to observe that in the first three and partly in the fourth step *one
has to put information into the process*. Information means data *and* know how from
people who work on the plant or have experience with the system under concern. In
the fourth and fifth step it is then possible to *extract information out of the process*.

3   The Preprocessing

Most data that are being used in industrial processes are noisy and erroneous. Usually, sensors are drifting, they may fall out, they often deliver strange values due to obscure reasons, they may be inconsistent and deliver values outside of the defined domain, etc. This means that usually the first step in data analysis is to remove those data that must be failures, which is a hard job in practice—one has to discuss with the operator or expert to check the ranges of all sensor inputs, including their expected means and standard deviations. Better than removing data items is to find reasonable substitutes by, e.g., filtering techniques so that one is not forced to remove valuable input/output vectors. Furthermore, filtering plays an important role in smoothing data. The comparison of expected ranges c.q. standard deviations with actual ones is an intensive job that can only partly be automated. The importance of *visualization* should be stressed here, it is important to produce pictures of two- or three-dimensional connections between variables.

3.1   The process data

The first data analysis step takes place in the sensors at the automation system on the plant. Here it is checked whether the sensors perform well, and a first filtering on the data takes place. The process data can, for example, be saved on a PC via the interface WIN-TM. At the same time the process data can be, for example, be registered in fixed interval times (e.g. 1 min.) on a PC in a dBase-file. In this way a data base with, e.g., the following process data comes into existence:

- Measurements (e.g., pressure, temperature, flow)

- Process information (information about local controllers, adjustments, opening widths, etc.)

- Machine adjustments (e.g., on/off)

- Quality data of the product

These quality data are divided into online and offline data. Offline data can be delivered by a laboratory that at fixed time intervals or process times checks the quality of the product from the plant. Besides the quality there can be offline data about the product amount, the kind of process that was used, and economical information about the plant like product losses, special events, hazards, and product sums. It is important to map the nonnumerical inputs on a numerical, possibly fuzzy scale, that takes into account the kind of offline data.

3.2   Cleaning the data

In the further process of data analysis each datum can be decisive for the resulting final model of the process. Therefore it is important to carefully observe the measured data with regard to sensor failures and runaways. First of all it is necessary to define the interval ranges of each sensor. This means that from each sensor it has to be known its upper and lower value and, if possible, its expected mean value and standard deviation. Only with this information it is possible to distinguish regular from irregular inputs and, if the process data are available on a time scala, to substitute expected input data for faulty data using a (fuzzy) filtering algorithm or another intelligent device. The following actions should be carried out:

- Remove columns of constants from the data set, or substitute them by expected values using a nonlinear or linear filter.

- Remove lines out of the data set that contain at least one value described by NaN (Not a Number). Here again, if filtering is possible and useful, substitute the NaN by an assumed and realistic value. The same applies to the case where at least one measured value is situated outside of the predefined measurement interval.

To mark measurements as NaN can be performed automatically, but has usually to be performed manually using an editor. In some projects within Siemens we used special drift recognition programs to observe how the data behave with regard to the predefined expected mean values of the sensors.

3.3   Normalizing the data

As each sensor and other inputs are defined on domains that are typical to their physical behavior it is useful to map them onto normalized domains, e.g., $[-1, 1]$ or $[0, 1]$ (see Fig. 1 with the interval $[-4, 4]$). This should prevent the fact that sensors defined on numerical larger domains than other ones get higher weights in the further data processing. The mathematical mapping of the standard deviation on the normalized domain should be regarded.

Babuška [2] has shown in his PhD-thesis that clustering with the fuzzy c-means algorithm in a multi-dimensional environment produces completely different result for normalized and unnormalized data. The Gustafson-Kessel algorithm, however, is almost independent from normalized and unnormalized conditions.

4   Data Reduction

Although there are cases where almost no data are available and one has to check carefully how each data item can be used perhaps twice to learn or check the desired

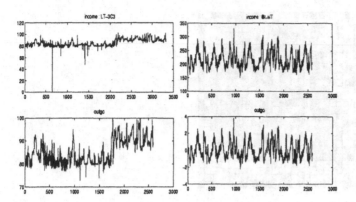

Figure 1: Left: the result of cleaning the data. Right: the result of normalizing the data.

model, usually it is the case that there are too many sensor inputs and enough data. An important question is which inputs are of relevance to the product quality of a plant or the diagnostic statement about a system. There are several methods that support this step. First, a correlation analysis gives information about the influence of inputs on the output, although one has to be careful in weighting correlation tests, they can only find a linear correspondence between variables. Secondly, a ranking test gives information about the importance of each parameter, in particular when we mix some random values amongst the input parameters. Thirdly, it is important to carry out a principal component test based on the correlation analysis. This usually greatly supports the reduction of the dimensions and gives a good insight in the importance of sensor inputs.

We will now in a couple of sections deal more closely with the above mentioned steps. A complete description can be found in [1].

4.1   Correlation analysis

There are several ways to deal with correlation coefficients. The first step should be to prepare a matrix with all mutual correlation values (cf. Fig. 2). This can be done via the covariance matrix. In systems with several hundreds of inputs and outputs it often turns out that there are large clusters of sensors that are mutually correlated with values of 0.9 or even 0.99, e.g., in the case that temperatures are calculated at several places along the plant. A correlation matrix can be a helpful means to reduce the number of inputs/outputs that have to be considered for further study of causal relationships between variables.

One can also take one or several important output parameters and calculate the

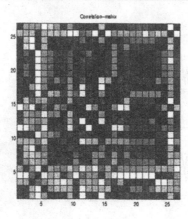

Figure 2: The correlation matrix.

correlations of all inputs with this output. This is useful if one wants to find out which inputs have highest influence on one of the outputs, e.g., the quality of the product (Fig. 3). These inputs will play an important role in later data analysis steps. Usually, only a small number of inputs have really significant correlation values with these outputs, i.e., one can considerably reduce the number of variables to be regarded more closely with this method.

The disadvantage of the correlation method is obvious for anyone who has been working with this method. If the data points of two inputs are not on a straight line, but, e.g., in a circle or on some very curvy line, the correlation value, a one-dimensional scalar, does not contain any valuable information. We present three different ways to uncover important relationships between input and output values, viz. the ranking test, PCA, and polynomial regression.

4.2   Ranking

This methods presumes the existence of an RBF-model, a model of the relationships between inputs and outputs via a neural network with radial basis functions. If such a modeling exists one can substitute each variable consecutivly by its mean value. Then one has to calculate for each variable the difference between the error of the RBF-model with the mean value taken as input and the model with the actual value of this input, i.e.,

$$diff(x_i) = E(\ldots, \bar{x}_i, \ldots) - E(\ldots, x_i, \ldots) \tag{1}$$

One has to sum up all $diff(x_i)$ to achieve one scalar difference value. One can now compare these error values for each input to obtain a ranking of the importance of each variable (see Fig. 4).

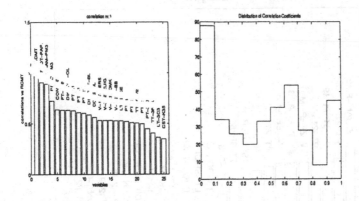

Figure 3: RCMT has correlation 1 with itself. Then follow other parameters.

One has to be careful in the interpretation of this value. It is certainly true that those inputs that have high differences are influential parameters in the process. On the other hand, however, if two or more variables are highly correlated (an event that often occurs), than removing one variable (substituting it by its mean value) will not result in any significant difference in the error values. Hence low values in the ranking test should be considered with care. Actually, this method cannot be used without the correlation test presented above.

4.3   Principal component analysis

Principal component analysis (PCA) is a very well known method that has been described in many places in literature. The user presents the number n of orthogonal axes that have to be obtained. Usually, n is 2 or 3, although this small number means the loss of a lot of valuable information. The resulting space is usually hard to interpret because the axes tend to be complex combinations of the original axes, depending on the data structure. The process takes the following steps: (1) Normalize the data, (2) Save the data in a matrix X, (3) Calculate the square matrix $A = X^T X$, and (4) Calculate the eigenvalues and eigenvectors of A. In a normalized data domain the size of the eigenvalue shows the importance of the attached eigenvector (cf. Fig. 5). Therefore, one should take the two or three most important eigenvectors as a new base of á reduced space frame. Furthermore, one should observe which inputs (considerably) contribute to these eigenvectors and to which degree. These inputs turn out to be the most influential parameters in the process. Hence, in an indirect way, this is also a useful method to determine the most important or most influential parameters of the process.

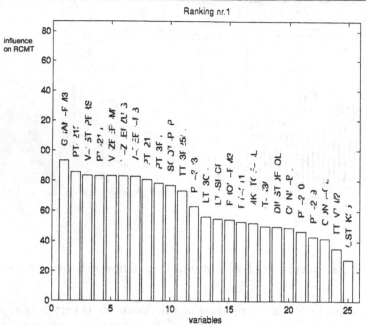

Figure 4: A ranking of several parameters.

4.4 Polynomial regression

Polynomial regression can be regarded as an intelligent correlation method. The correlation method attempts to put a straight line through the data points of two inputs or outputs. With polynomial regression it is possible to generate a polynomial curve that approximates the data points (Fig. 6). In practice, polynomials of degree 3 or 4 are the highest values one should take. More complex relationships are too hard to understand for the user.

5   Modeling

Based on the available data a model should be made that contains the dependencies between variables The easiest and surely not the worst way is to build a radial basis function (RBF) network. Usually, a simple RBF-network has a good performance in representing the input-output connection.

Another method that is frequently being used is fuzzy clustering [3]. There is fuzzy clustering of the input domain, of the output domain, of the relation between a number of inputs and outputs, etc. Furthermore, there is point clustering, line clustering, and also clustering around more complex structures like *c-elliptotypes* [11].

Thirdly, to gain insight into the model one can use Kohonen or similar mappings, e.g., the Sammon mapping. These mappings show the structure of the data in a two- or three-dimensional figure, and help to show the timely behavior of processes visually, which is important, as sensors as well as plants tend to drift during the time. Process operators are particularly interested to observe the state of their process during the time.

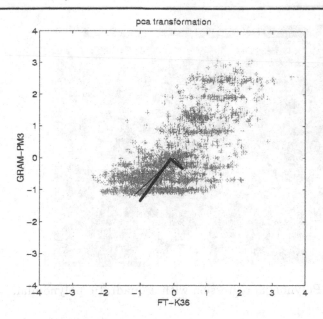

Figure 5: The location of two axes after principal component analysis.

5.1 RBF-networks

It is not our purpose to give a detailed description of neural networks here. We will only deal with a special kind of neural networks, viz. networks with radial basis functions. Radial basis functions are usually multi-dimensional Gauß-functions. RBF-networks mainly play a role as cluster algorithms in a supervised learning environment. In a three layer neural network of this kind the network output y is equal to

$$y = NN(\mathbf{x}) = \frac{\sum_i w_i(\mathbf{x}) b_i(\mathbf{x})}{\sum_i b_i(\mathbf{x})} \qquad (2)$$

w_i can be regarded as the output of a fuzzy rule; b_i as the input of a fuzzy rule, b_i can be any function but is usually chosen as the Gauß-function

$$b_i(x) = \kappa_j \exp\left(-\sum_{j=1}^{L} \frac{(x_j - c_{ij})^2}{2\sigma_{ij}^2}\right) \qquad (3)$$

Figure 7 shows the results of training an RBF-network. It can be easily shown that RBF-networks can be mapped on a special class of fuzzy systems and vice versa. Martinetz and Hollatz [9] present the following conditions:

1. The number of basis functions is equal to the number of fuzzy if-then rules.

2. The fuzzy rules are singleton or Takagi-Sugeno rules.

3. The membership functions in the fuzzy system are Gauß-functions.

4. The and-operation in the fuzzy system is performed by multiplication.

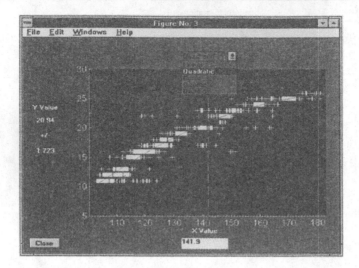

Figure 6: Polynomial regression with a quadratic polynomial.

5. Both the RBF-network and the fuzzy system use the same defuzzification method.

An RBF-network that satisfies these criteria can easily be mapped into a fuzzy system, which allows the interpretation of the contents of the neural network. Martinetz and Hollatz [9] used this system on a steel rolling mill. This resulted in rules like:

1. **If** b is *low*, T is *high*, and ϵ is *medium* f **then** decrease the rolling force by -28.9 MN.

2. **If** V is *medium*, b is *high*, and ϵ is *high* bf **then** increase the rolling force by 14.7 MN.

3. **If** d is *high* **then** decrease the rolling force by -6.8 MN.

4. **If** b is *high*, T is *low*, and ϵ is *high* **then** increase the rolling force by 12.6 MN.

Hence RBF-networks are one way to model the process and to obtain information out of process data.

5.2 Fuzzy clustering

Fuzzy clustering has been described in many papers and books [3], most recently in the dissertation of Babuška [2]. Fuzzy clustering is somewhat similar to RBF-networks, it takes the data and learns out of the data points a fixed number of fuzzy classes. Each point may belong to a certain degree to one or more classes. These classes may be centered around a point, but also, more interestingly, along a straight line or even along an elliptotype. When a fuzzy clustering exists, one can easily obtain fuzzy rules out of these clusters.

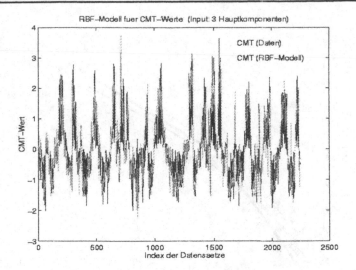

Figure 7: The result of training an RBF-network.

The cluster centers and the membership degree of data to each cluster center are calculated in an iterative process using

$$v_i = \frac{\sum_{k=1}^{n} u_{ik}^m x_k}{\sum_{k=1}^{n} u_{ik}^m} \tag{4}$$

where v_i are the cluster centers, x_k the data points, and

$$u_{ik} = \sum_{j=1}^{c} \left(\frac{d_{ik}}{d_{jk}} \right)^{-\frac{2}{m-1}} \tag{5}$$

is the membership degree to cluster i, and d is a distance function. This may be the distance to a point as well as the distance to a line (Fig. 8) or a more complicated mathematical function. In the elliptotype-case the distance function becomes fairly complicated, as one can imagine.

The order of the data points given into the iteration procedure influences the location and size of the clusters.

5.3   Kohonen mapping

In [8] Kohonen proposed a method to use unsupervised neural network learning to map an n-dimensional space-frame to, e.g., a 3-dimensional (Fig. 9) or a 2-dimensional (Fig. 10) mapping, preserving as much as possible the structure in the data as well as the distances between neighboring data points. Therefore one needs an appropriate neighboring function. The visualization from 2-dimensional Kohonen mappings may give some insight into the structure of the data.

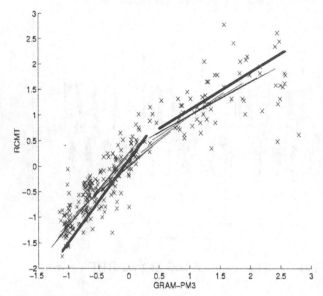

Figure 8: The resulting lines after fuzzy clustering.

It is particularly interesting to focus on one crucial output value. If one produces several maps with different clustered input variables and the same output one can compare the relationship between these variables and the output. In some cases it is possible with this method to save expensive chemicals and achieve good quality output. Of course, this presumes know how from experts from the plant. In a real life project we had some very interesting results in this area, that led to interesting savings for the plant owner.

5.4   Sammon mapping

A Sammon mapping [5] (cf. Fig. 11) has the same goal as a Kohonen mapping but uses a different way to achieve this goal. Kohonen uses a neural network in which all connections between variables are summarized, in a Sammon mapping gradient descent methods are used to map the n-dimensional space onto an $(n-1)$-dimensional and so on until the 2-dimensional space is reached. Goal is to preserve as much as possible the Euclidean distances between the data points. The choice of the energy function in this process is crucial.

There are the following steps:

1. There is a n-dimensional space and a corresponding 2- (or 3-)dimensional space.

2. There are distances d in the n-dimensional space frame and corresponding distances d^* in the 2-dimensional space.

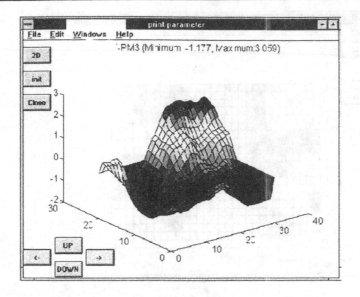

Figure 9: A three dimensional Kohonen mapping result.

3. Calculate

$$E(m) = \frac{1}{\sum_{i<j} d_{ij}^*} \sum_{i<j}^{N} \frac{(d_{ij}^* - d_{ij}(m))^2}{d_{ij}^*} \qquad (6)$$

a measure for the differences in the distances in the two spaces after the m-th iteration cycle.

4. The adaptation rule in the m-the iteration cycle is given by a kind of linearization process, viz. $y_{pq}(m+1) = y_{pq}(m) - \alpha \cdot \Delta_{pq}(m)$. α is the adaptation rate that has to be determined empirically, Δ is given by

$$\Delta_{pq}(m) = \frac{\partial E(m)/\partial y_{pq}(m)}{\left| \partial^2 E(m)/\partial y_{pq}(m)^2 \right|}. \qquad (7)$$

6   Rule extraction

Another step is to extract information in the form of fuzzy rules out of the process or system. There are several ways to do this. One is via RBF-networks, it is not so hard to translate an RBF-network into a fuzzy system. A disadvantage of this approach is the complexity of the rules that arises. Each rule has its own premises with newly defined fuzzy sets, which makes it hard to understand the meaning of the rules. Another method is building rules by clustering techniques. The fuzzy clustering of input data, for example, corresponds to the input fuzzy sets. Using this method delivers rules that are usually of great interest to the customer. In some cases these rules immediately led to important savings, because relations that were not known until that time became

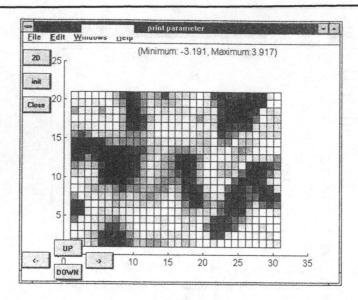

Figure 10: A two dimensional Kohonen mapping result.

apparent. Figures 12 and 13 show some screen dumps of our tool after RBF-training and fuzzy clustering.

6.1 Rule extraction from RBF-networks

We have seen before how the extraction of rule from RBF-networks proceeds. We have to point to a problem, however, that in cases of many inputs, a PCA is needed to reduce the number of inputs. This PCA then produces, say, three main axes, that are usually hard to interprete linguistically. Rules produced on the basis of these axes are hard to understand for the people from the plant. Therefore, one should carefully explain them the rules. If one is able to determine a small number of really important inputs, it is better to use these inputs for RBF-modelling and rule production. This results in rules with understandable premises, the conclusions remain crisp, usually very precise values.

6.2 Rule extraction from fuzzy clustering

Fuzzy clustering is an interesting method to produce rules, because one can derive a number of local linear centers that can be combined via Takagi-Sugeno rules [4]. This means that the Takagi-Sugeno system performs an interpolation between the cluster centers. As this interpolation is performed inside of the system, the rules remain understandable. Produced rules in one of our projects had the following form:

> If $Input_1$ is A_1 then Output = $1.41955 \times Input_1 + 0.170627$.
> If $Input_1$ is A_2 then Output = $0.536239 \times Input_1 + 0.52224$.

The two clusters A_1 and A_2 represent two important domains of $Input_1$.

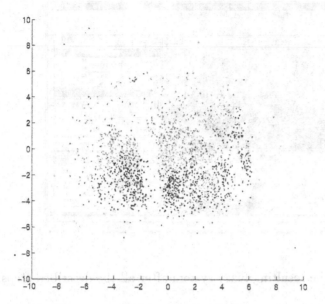

Figure 11: The result of a Sammon mapping.

7 Drift Analysis

Drift analysis is easy when one knows the natural state of the plant. In this case, one has to describe this state and calculate the difference between the current state and the predefined state. In many cases, however, one does not know the standard state of a plant or a system, in this case one has to follow carefully the movements in the data. If there are jumps in several variables that are abnormal with respect to time, one should give a signal. This usually implies that one has to do a reidentification. This reidentification is dangerous, because one has to know which data to chose for this process. If one uses data that is too local with respect to time, one might be learning a system that has nothing in common with the system as a whole. If a new drift occurs, the system cannot be controlled appropriately. A solution is to save some standard data and use them as extra data to reidentify the system. This is dangerous too, because systems have natural drifts and might never recur to such a 'standard' state. Careful analysis is needed in this case.

8 Application Areas

We briefly present some projects out of a larger pool of projects within Siemens (cf., e.g., [6]) that were concerned with data analysis. In our talk we will more extensively deal with these projects.

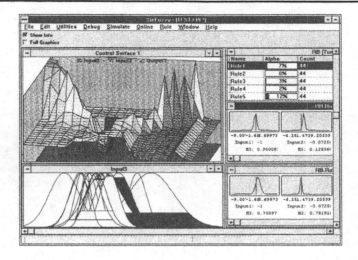

Figure 12: The resulting fuzzy system after all the described steps.

8.1 Traffic Control

In many large cities systems are implemented that tell a driver when he enters the
city about the situation of the parking garages downtown. Usually this is done by
signs with information about the current number of free places in the parking houses
downtown. A better concept is to have signs that forecast the situation of a parking
garage after the estimated driving time towards the garage. To make such a prognosis
one needs a lot of knowledge about the estimated traffic situation in the city and about
the expected number of other cars also driving to the parking garage. We have build a
fuzzy system that gives such a forecast. After that, we used data from several parking
garages, weather data, and general traffic data and processed them in a neural network.
This highly improved the quality of the fuzzy system [7].

Another application in traffic control is taking place in Cologne, Germany. From
120 inductions loops we gather information about the traffic flow in a part of the city.
These data plus information about special events like fairs, sporting events, etc. are
analyzed and used to deliver a daily prognosis message in the newspaper and an hourly
traffic report on the local television.

8.2 Paper Industry

In one of the many projects in paper industry we had to deal with a recycling process,
with numerous data describing this process. In a compact form we dealt with 3,300
data vectors, each consisting of 256 sensor values. We used these data to model the
plant, to show the most important sensor values, to show drifting of the plant, and to
represent knowledge about the behavior of the plant in the form of fuzzy rules.

In another project in pulp processing we had to tell which of the sensors that were

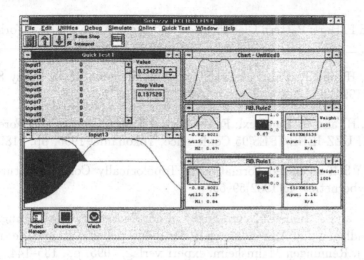

Figure 13: Some variables of the final fuzzy system.

being tested were of importance to the quality of the process. Lack of good data and steady changes in the process made the analysis more difficult than expected. Nevertheless, some important statements about the choice of the sensors were of great help to the company.

8.3   Automotive Industry

In car industry we had a test driver driving a car with an installed camera behind the front window. We obtained data from a number of existing conventional and electronic components like ABS, power steering, engine control, etc. and had to extract the driving situation out of these data. This could be done by the above mentioned steps. The derived fuzzy system was translated to a neural network, improved by the data, and translated back into a fuzzy system.

References

[1] Ader, W. and A. Nörling: Extraktion von Fuzzy-Regeln zur Analyse technischer Prozesse, MSc-thesis, Munich, 1995.

[2] Babuška, R.: Fuzzy Modeling and Identification, PhD-thesis, Delft, 1996.

[3] Bezdek, J.C.: Pattern Recognition with Fuzzy Objective Function Algorithms, New York, Plenum Press, 1981.

[4] Driankov, D., H. Hellendoorn and M. Reinfrank: An Introduction to Fuzzy Control, Heidelberg, Springer Verlag, 1993.

[5] Duda, R. and P. Hart: Pattern Classification and Scene Analysis, New York, Wiley, 1973.

[6] Hellendoorn, H. and R. Palm: Fuzzy Systems at Siemens R&D, Fuzzy Sets and Systems, 63(1994)245–269.

[7] Hellendoorn, H. and R. Baudrexl: Fuzzy-Neural Traffic Control and Forecasting, Proc. of the FUZZ-IEEE/IFES'95 Conference, Yokohama, 1995, pp. 2187–2194.

[8] Kohonen, T.: Self-organized Formation of Topologically Correct Feature Maps, Biological Cybernetics, 43(1982)59–69.

[9] Martinetz, T. and J. Hollatz: Neuro-Fuzzy in der Prozeßautomatisierung, Neuro-Fuzzy – Grundlagen und Anwendungen in der industriellen Automatisierung, (Ed. K.W. Bonfig), Renningen-Malmsheim, expert verlag, 1995, pp. 135–144.

[10] Palm, R., D. Driankov, and H. Hellendoorn: Model Based Fuzzy Control, Heidelberg, Springer Verlag, 1997.

[11] Runkler, T.: Iterative c-Elliptotype Clustering for Efficient Data Analysis, Workshop Fuzzy-Neuro-Systeme (Ed. R. Isermann), Darmstadt, 1995.

FUZZY SHELL CLUSTER ANALYSIS

F. Klawonn, R. Kruse and H. Timm
University of Magdeburg, Magdeburg, Germany

Abstract

In this paper we survey the main approaches to fuzzy shell cluster analysis which is simply a generalization of fuzzy cluster analysis to shell like clusters, i.e. clusters that lie in nonlinear subspaces. Therefore we introduce the main principles of fuzzy cluster analysis first. In the following we present some fuzzy shell clustering algorithms. In many applications it is necessary to determine the number of clusters as well as the classification of the data set. Subsequently therefore we review the main ideas of unsupervised fuzzy shell cluster analysis. Finally we present an application of unsupervised fuzzy shell cluster analysis in computer vision.

1 Introduction

Cluster analysis is a technique for classifying data, i.e. to divide the given data into a set of classes or *clusters*. In classical cluster analysis each datum has to be assigned to exactly one class. Fuzzy cluster analysis relaxes this requirement by allowing gradual memberships, offering the opportunity to deal with data that belong to more than one class at the same time.

Traditionally, fuzzy clustering algorithms were used to search for compact clusters. Another approach is to search for clusters that represent nonlinear subspaces, for instance spheres or ellipsoids. This is done using fuzzy shell clustering algorithms, which is the subject of this paper.

Fuzzy shell cluster analysis is based on fuzzy cluster analysis. Therefore we review the main ideas of fuzzy cluster analysis first, and present then some fuzzy shell clustering algorithms. These algorithms search for clusters of different shapes, for instance ellipses, quadrics, ellipsoids etc. Since in many applications the number of clusters, into which the data shall be divided, is not known in advance, subsequently the subject of unsupervised fuzzy shell clustering analysis is reviewed. Unsupervised fuzzy shell clustering algorithms determine the number of clusters as well as the classification of the data set. Finally an application of fuzzy shell cluster analysis in computer vision is presented.

2   Fuzzy Cluster Analysis

2.1   Objective Function Based Clustering

Objective function based clustering methods determine an optimal classification of data by minimizing an objective function. Depending on whether binary or gradual memberships are used, one distinguishes between hard and fuzzy clustering methods. In fuzzy cluster analysis data can belong to several clusters at different degrees and not only to one. In general the performance of fuzzy clustering algorithms is superior to that of the corresponding hard algorithms [1].

In objective function based clustering algorithms each cluster is usually represented by a prototype. Hence the problem of dividing a data set X, $X = \{x_1, \ldots, x_n\} \subseteq \mathbb{R}^p$, into c clusters can be stated as the task of minimizing the distances of the datum to the prototypes. This is done by minimizing the following objective function $J(X, U, \beta)$

$$J(X, U, \beta) = \sum_{i=1}^{c} \sum_{j=1}^{n} u_{ij}^m d^2(\beta_i, x_j) \tag{1}$$

subject to

$$\sum_{j=1}^{n} u_{ij} > 0 \quad \text{for all } i \in \{1, \ldots, c\} \tag{2}$$

$$\sum_{i=1}^{c} u_{ij} = 1 \quad \text{for all } j \in \{1, \ldots, n\} \tag{3}$$

where $u_{ij} \in [0, 1]$ is the membership degree of datum x_j to cluster i, β_i is the prototype of cluster i, and $d(\beta_i, x_j)$ is the distance between datum x_j and prototype β_i. The $c \times n$ matrix $U = [u_{ij}]$ is also called the fuzzy partition matrix and the parameter m is called the fuzzifier. Usually $m = 2$ is chosen.

Constraint (2) guarantees that no cluster is empty and constraint (3) ensures that the sum of membership degrees for each datum equals 1. Fuzzy clustering algorithms which satisfy these constraints are also called *probabilistic clustering algorithms*, since the membership degrees for one datum formally resemble the probabilities of its being a member of the corresponding cluster.

The objective function $J(X, U, \beta)$ is usually minimized by updating the membership degrees u_{ij} and the prototypes β_i in an alternating fashion, until the change ΔU of the membership degrees is less than a given tolerance ε. This approach is also known as the alternating optimization method.

A Fuzzy Clustering Algorithm
Fix the number of clusters c
Fix m, $m \in (1, \infty)$
Initialize the fuzzy c-partition U
REPEAT
    Update the parameters of each clusters prototype

Update the fuzzy c-partition U using (4)
UNTIL $|\Delta U| < \varepsilon$

To minimize the objective function (1), the membership degrees are updated using (4). The following equation for updating the membership degrees can be derived by differentiating the objective function (1).

$$
u_{ij} = \begin{cases} \dfrac{1}{\displaystyle\sum_{k=1}^{c}\left(\dfrac{d^2(x_j,\beta_i)}{d^2(x_j,\beta_k)}\right)^{\frac{1}{m-1}}} & \text{if } I_j = \emptyset, \\[4mm] 0 & \text{if } I_j \neq \emptyset \text{ and } i \notin I_j, \\[2mm] x, x \in [0,1] \text{ such that } \sum_{i\in I_j} u_{ij} = 1, & \text{if } I_j \neq \emptyset \text{ and } i \in I_j. \end{cases}
\tag{4}
$$

This equation is used for updating the membership degrees in every probabilistic clustering algorithm.

In contrast to the minimization of the objective function (1) the minimization of (1) varies with respect to the prototypes according to the choice of the prototypes and the distance measure. Therefore each choice leads to a different algorithm.

2.2   Possibilistic Clustering Algorithms

The prototypes are not always determined correctly using probabilistic clustering algorithms, i.e. only a suboptimal solution is found. The main source of the problem is constraint (3), which requires the membership degrees of a point across all clusters to sum up to 1. This is easily demonstrated by considering the case of two clusters. A datum x_1, which is typical for both clusters, has the same membership degrees as a datum x_2, which is not at all typical for any of them. For both data the membership degrees are $u_{ij} = 0.5$ for $i = 1, 2$. Therefore both data influence the updating of the clusters to the same extent.

An obvious modification is to drop constraint (3). To avoid the trivial solution, i.e. $u_{ij} = 0$ for all $i \in \{1, \ldots, c\}, j \in \{1, \ldots, n\}$, (1) is modified to (5).

$$
J(X, U, \beta) = \sum_{i=1}^{c}\sum_{j=1}^{n} u_{ij}^m d^2(\beta_i, x_j) + \sum_{i=1}^{c} \eta_i \sum_{j=1}^{n}(1 - u_{ij})^m
\tag{5}
$$

where $\eta_i > 0$.

The first term minimizes the weighted distances while the second term avoids the trivial solution. A fuzzy clustering algorithm that minimizes the objective function (5) under the constraint (2) is called a *possibilistic clustering algorithm*, since the membership degrees for one datum resemble the possibility of its being a member of the corresponding cluster.

Minimizing the objective function (5) with respect to the membership degrees leads to the following equation for updating the membership degrees u_{ij} [11].

$$u_{ij} = \frac{1}{1 + \left(\dfrac{d^2(x_j, \beta_i)}{\eta_i}\right)^{\frac{1}{m-1}}} \tag{6}$$

Equation (6) shows, that η_i determines the distance at which the membership degree equals 0.5. If $d^2(x_j, \beta_i)$ equals η_i, the membership degree equals 0.5. So it is useful, to choose η_i for each cluster separately [11]. η_i can be determined by using the fuzzy intra cluster distance (7) for example.

$$\eta_i = \frac{K}{N_i} \sum_{j=1}^{n} (u_{ij})^m d^2(x_j, \beta_i). \tag{7}$$

where $N_i = \sum_{j=1}^{n}(u_{ij})^m$. Usually $K = 1$ is chosen.

It is recommended to initialize a possibilistic clustering algorithm with the results of the corresponding probabilistic version [12]. In case prior information about the clusters is available, it can be used to determine η_i for a further iteration of the fuzzy clustering algorithm to fine tune the results [10].

A Possibilistic Clustering Algorithm
Fix the number of clusters c
Fix m, $m \in (1, \infty)$ Initialize U using the corresponding fuzzy algorithm
Compute η_i using (7)
REPEAT
    Update prototype using U
    Compute U using (6)
UNTIL $|\Delta U| < \varepsilon_1$
Fix the values of η_i using a priori information ⎫
REPEAT   ⎪
    Update prototype using U   ⎬ *optional*
    Compute U using (6)   ⎪
UNTIL $|\Delta U| < \varepsilon_2$ ⎭

2.3   The Fuzzy C Means Algorithm

The simplest fuzzy clustering algorithm is the *fuzzy c means algorithm (FCM)* [1]. The c in the name of the algorithm reminds that the data is divided into c clusters. The FCM searches for compact clusters which have approximately the same size and shape. Therefore the prototype is a single point which is the center of the cluster, i.e. $\beta_i = (c_i)$. The size and shape of the clusters are determined by a positive definite $n \times n$ matrix A. Using this matrix A the distance of a point x_j to the prototype β_i is given by

$$d^2(x_j, \beta_i) = \|x_j - c_i\|_A^2 = (x_j - c_i)^T A(x_j - c_i). \tag{8}$$

In case A is the identity matrix, the FCM looks for spherical clusters otherwise for ellipsoidal ones. In most cases the Euclidean norm is used, i.e. A is the identity matrix. Hence the distance reduces to the Euclidean norm, i.e.

$$d^2(x_j, \beta_i) = \|x_j - c_i\|^2. \qquad (9)$$

Minimizing the objective function with respect to the prototypes leads to the following equation (10) for updating the prototypes [7].

$$c_i = \frac{1}{N_i} \sum_{j=1}^{n} (u_{ij})^m x_j \qquad (10)$$

where $N_i = \sum_{j=1}^{n} (u_{ij})^m$.

A disadvantage of the FCM is, that A is not updated. Therefore the shape of the clusters cannot be changed. Besides, when the clusters are of different shape, it is not appropriate to use a single matrix A for all clusters at the same time.

2.4 The Gustafson-Kessel Algorithm

The *Gustafson-Kessel algorithm (GK)* searches for ellipsoidal clusters [6]. In contrast to the FCM, a separate matrix A_i, $A_i = (\det C_i)^{1/n} C_i^{-1}$, is used for each cluster. The norm matrices are updated as well as the centers of the corresponding clusters. Therefore the prototypes of the clusters are a pair (c_i, C_i), where c_i is the center of the cluster and C_i the covariance matrix, which defines the shape of the cluster.

Like the FCM the GK computes the distance to the prototypes by

$$d^2(x_j, \beta_i) = (\det C_i)^{1/n} (x_j - c_i)^T C_i^{-1} (x_j - c_i). \qquad (11)$$

To minimize the objective function with respect to the prototypes, the prototypes are updated according to the following equations [7]

$$c_i = \frac{1}{N_i} \sum_{j=1}^{n} (u_{ij})^m x_j, \qquad (12)$$

$$C_i = \frac{1}{N_i} \sum_{j=1}^{n} (u_{ij})^m (x_j - c_i)(x_j - c_i)^T. \qquad (13)$$

The GK is a simple fuzzy clustering algorithm to detect ellipsoidal clusters with approximately the same size but different shapes. In combination with the FCM it is often used to initialize other fuzzy clustering algorithms. Besides the GK can also be used to detect linear clusters. This is possible, because lines and planes can also be seen as degenerated ellipses or ellipsoids, i.e. at least in one dimension the radius nearly equals zero.

2.5 Other Algorithms

There are many fuzzy clustering algorithms besides the FCM and the GK. These algorithms search for clusters with different shape, size and density of data and use different distance measures. For example, if one is interested in ellipsoidal clusters of varying size the Gath and Geva algorithm can be used [5]. It searches for ellipsoidal clusters, which can have different shape, size, and density of data.

If one is interested in linear clusters, for instance lines, linear clustering algorithms, for example the fuzzy c-varieties algorithm [1] or the adaptive fuzzy clustering algorithm [3], can be used. Another linear clustering algorithm is the compatible cluster merging algorithm (CCM) [8, 7]. This algorithm uses the property of the GK to detect linear clusters and improves the results obtained by the GK by merging compatible clusters. Two clusters are considered compatible, if the distance between these clusters is small compared to their size and if they lie in the same hyperplane.

A common application of the CCM is line detection. The advantage of the CCM in comparison to other line detection algorithms is its ability to detect significant structures while neglecting insignificant ones.

3   Fuzzy Shell Cluster Analysis

The fuzzy clustering algorithms discussed up to now search for clusters that lie in linear subspaces. Besides, it is also possible to detect clusters that lie in nonlinear subspaces, i.e. resemble shells or patches of surfaces with no interior points. These clusters can be detected using fuzzy shell clustering algorithms.

The only difference between fuzzy clustering algorithms and fuzzy shell clustering algorithms is that the prototypes of fuzzy shell clustering algorithms resemble curves resp. surfaces or hypersurfaces. Therefore the algorithm for probabilistic clustering and the algorithm for possibilistic clustering are both used for fuzzy shell cluster analysis.

There is a large number of fuzzy shell clustering algorithms which use different kinds of prototypes and different distance measures. Fuzzy shell clustering algorithms can detect ellipses, quadrics, polygons, ellipsoids, hyperquadrics etc. In the following the fuzzy c ellipsoidal shells algorithm, which searches for ellipsoidal clusters, and the fuzzy c quadric shells algorithm, which searches for quadrics, are presented. Further fuzzy shell clustering algorithms are described in [7].

3.1   The Fuzzy C Ellipsoidal Shells Algorithm

The *fuzzy c ellipsoidal shells algorithm (FCES)* searches for shell clusters with the shape of ellipses, ellipsoids or hyperellipsoids [7, 4]. In the following we present the algorithm to find ellipses.

An ellipse is given by

$$(x - c_i)^T A_i (x - c_i) = 1, \tag{14}$$

where c_i is the center of the ellipse and A_i is a positive symmetric matrix, which determines the major and minor axes lengths as well as the orientation of the ellipse. From that description of an ellipse the prototypes β_i, $\beta_i = (c_i, A_i)$, for the clusters are derived.

The fuzzy c ellipsoidal shells algorithm uses the radial distance. This distance measure is a good approximation to the exact (perpendicular) distance, but easier to compute. The radial distance d^2_{Rij} of a point x_j to a prototype β_i is given by

$$d^2(x_j, \beta_i) = d^2_{Rij} = \|x_j - z\|^2, \tag{15}$$

where z is the point of the intersection of the ellipse β_i and the line through c_i and x_j that is near to the cluster.

Using (14) d^2_{Rij} can be transformed to

$$d^2_{Rij} = \frac{\left(\sqrt{(x_j - c_i)^T A_i(x_j - c_i)} - 1\right)^2 \|x_j - c_i\|^2}{(x_j - c_i)^T A_i(x_j - c_i)}. \tag{16}$$

Minimizing the objective function with respect to the prototypes leads to the following system of equations [7]:

$$\sum_{j=1}^{n} u_{ij}^m (x_j - c_i)(x_j - c_i)^T \left(\frac{\|x_j - c_i\|}{d_{ij}}\right)^2 \left(\sqrt{d_{ij}} - 1\right) = 0, \tag{17}$$

$$\sum_{j=1}^{n} \frac{u_{ij}^m \left(\sqrt{d_{ij}} - 1\right)}{d_{ij}^2} \cdot \left[\|x_j - c_i\|^2 A_i + \left(\sqrt{d_{ij}} - 1\right) d_{ij} I\right](x_j - c_i) = 0, \tag{18}$$

where $d_{ij}^2 = (x_j - c_i)^T A_i(x_j - c_i)$ and I is the identity matrix.

This system of equations has to be solved using numerical techniques. To update the prototypes e.g. the Levenberg-Marquardt algorithm [13] can be used.

3.2   The Fuzzy C Quadric Shells Algorithm

The fuzzy c quadric shells algorithm ($FCQS$) searches for clusters with the shape of a quadric or a hyperquadric. A quadric resp. a hyperquadric is defined by

$$p_i^T q = 0, \tag{19}$$

where

$p_i^T = (p_{i1}, p_{i2}, \ldots, p_{in}, p_{i(n+1)}, \ldots, p_{ir}, p_{ir+1}, \ldots, p_{is}),$
$q^T = (x_1^2, x_2^2, \ldots, x_n^2, x_1 x_2, \ldots, x_{n-1} x_n, \ldots, x_1, x_2, \ldots, x_n, 1),$
$s = n(n+1)/2 + n + 1 = r + n + 1,$
n is the dimension of the feature vector of a datum and $r = n(n+1)/2$.
Hence the prototypes of the fuzzy c quadric shell clustering algorithm are s-tuples.

The FCQS uses the algebraic distance. The algebraic distance of a point x_j to a prototype β_i is defined by

$$d^2(x_j, \beta_i) = d^2_{Qij} = p_i^T q_j q_j^T p_i = p_i^T M_j p_i, \tag{20}$$

where $M_j = q_j q_j^T$.

An additional constraint is needed to avoid the trivial solution $p_i^T = (0, \ldots, 0)$. For two dimensional data the constraint

$$\|p_{i1}^2 + p_{i2}^2 + \ldots + p_{in}^2 + \frac{1}{2} p_{i(n+1)}^2 + \ldots + \frac{1}{2} p_{ir}^2\|^2 = 1 \tag{21}$$

is recommended, because it is a good compromise between performance and result quality [10]. However this constraint prevents the algorithm from finding linear clusters. Linear clusters are detected as hyperbolas or ellipses with a large ratio of major to minor axis. Therefore an additional algorithm for line detection is needed, which is executed after the FCQS. For that purpose the CCM is well suited. Good results are obtained by initializing the CCM, using those clusters, which probably represent linear clusters, i.e. hyperbolas and ellipses with a large ratio of major to minor axis [10].

Defining $a_i = (a_{i1}, \ldots, a_{in})$, $b_i = (b_{i1}, \ldots, b_{in})$ by

$$a_{ik} = \begin{cases} p_{ik} & 1 \leq k \leq n \\ \dfrac{p_{ik}}{\sqrt{2}} & n+1 \leq k \leq r \end{cases} \tag{22}$$

$$b_{ik} = p_{i(r+k)} \quad 1 \leq k \leq s - r \tag{23}$$

constraint (21) simplifies to $\|a_i\|^2 = 1$. To minimize the objective function with respect to the prototypes, a_i and b_i are computed by
a_i = eigenvector corresponding to the smallest eigenvalue of $(F_i - G_i^T H_i^{-1} G_i)$,
$b_i = -H_i^{-1} G_i a_i$,
where

$$F_i = \sum_{j=1}^n u_{ij}^m R_j, \quad G_i = \sum_{j=1}^n u_{ij}^m S_j, \quad H_i = \sum_{j=1}^n u_{ij}^m T_j,$$
$$R_j = r_j r_j^T, \quad S_j = r_j t_j^T, \quad T_j = t_j t_j^T,$$

$r_j^T = [x_{j1}^2, x_{j2}^2, \ldots, x_{jn}^2, \sqrt{2} x_{j1} x_{j2}, \ldots, \sqrt{2} x_{jk} x_{jl}, \ldots, \sqrt{2} x_{jn-1} x_{jn}]$,
$t_j^T = [x_{j1}, x_{j2}, \ldots, x_{jn}, 1]$.

Therefore updating the prototypes reduces to an eigenvector problem of size $n(n+1)/2$, which is trivial. However the chosen distance measure d^2_{Qij} is highly nonlinear in nature and is sensitive to the position of a datum x_j with respect to the prototype β_i [10]. Therefore the membership degrees computed using the algebraic distance are not very meaningful. Depending on the data, this sometimes leads to bad results.

Since this problem of the FCQS is caused by the particular distance measure, the modified FCQS uses the shortest (perpendicular) distance d_{Pij}^2. To compute this distance, we first rewrite (19) as $x^T A_i x + x^T b_i + c_i = 0$. Then the shortest distance between a datum x_j and a cluster β_i is given by [10]

$$d^2(x_j, \beta_i) = d_{Pij}^2 = \min_z \|x_j - z\|^2, \tag{24}$$

subject to

$$z^T A_i z + z^T b_i + c_i = 0, \tag{25}$$

where z is a point on the quadric β_i. By using the Lagrange multiplier λ, the solution is found to be

$$z = \frac{1}{2}(I - \lambda A_i)^{-1}(\lambda b_i + 2x_j), \tag{26}$$

where I is the identity matrix. Substituting (26) in (25) yields a forth degree equation in λ. Each real root λ_k of this polynomial represents a possible value for λ. Calculating the corresponding z vector z_k, d_{Pij}^2 is determined by

$$d_{Pij}^2 = \min_k \|x_j - z_k\|^2. \tag{27}$$

The disadvantage of using the exact distance is, that the modified FCQS is computationally very expensive, because updating the prototypes can be achieved only by numeric techniques such as the Levenberg-Marquardt algorithm [13, 10, 4]. Therefore using a simplified modified FCQS is recommended. In this simplified algorithm the prototypes are updated using the algebraic distance d_{Qij} and the membership degrees are updated using the shortest distance d_{Pij} [10].

In higher dimensions the approximate distance d_{Aij} is used instead of the geometric distance d_{Pij}. It is defined by:

$$d^2(x_j, \beta_i) = d_{Aij}^2 = \frac{d_{Qij}^2}{|\nabla d_{Qij}|^2} = \frac{p_i^T M_j p_i}{p_i^T (D(q_j) D(q_j)^T) p_i} \tag{28}$$

where ∇d_{Qij} is the gradient of the functional $p_i^T q$ evaluated in x_j and $D(q_j)$ the Jacobian of q evaluated in x_j. The corresponding variant of the FCQS is called the fuzzy c planoquadric shells algorithm (FCPQS) [10].

The reason for using the approximate distance is that there is no closed form solution for d_{Pij} in higher dimensions. Hence in higher dimensions the modified FCQS cannot be applied.

Updating the prototypes of the FCPQS requires solving a generalized eigenvector problem, for instance on the basis of the QZ algorithm [10].

4   Unsupervised Fuzzy Shell Cluster Analysis

The algorithms discussed so far are based on the assumption that the number of clusters is known beforehand. However, in many applications the number of clusters c into which a data set shall be divided is not known.

This problem can be solved using unsupervised fuzzy clustering algorithms. These algorithms determine automatically the number of clusters by evaluating a computed classification on the basis of validity measures.

There are two kinds of validity measures, local and global. The former evaluates single clusters while the latter evaluates the whole classification. Depending on the validity measure, unsupervised fuzzy clustering algorithms are divided into algorithms based on local validity measures and algorithms based on global validity measures.

In this section the ideas of unsupervised fuzzy clustering are presented. A detailed discussion can be found in [7].

4.1   Global Validity Measures

An unsupervised fuzzy clustering algorithm based on a global validity measure is executed several times, each time with a different number of clusters. After each execution the clustering of the data set is evaluated. Since global validity measures evaluate the clustering of a data set as a whole, only a single value is computed. Usually the number of clusters is increased until the evaluation of the clustering indicates that the solution becomes worse.

However it is very difficult to detect a probably optimal solution as is easily demonstrated. A very simple global validity measure is the objective function of the fuzzy clustering algorithm. But it is obvious that the global minimum of that validity measure is unusable, because the global minimum is reached, if the number of data equals the number of clusters. Therefore often the apex of the validity function is used instead.

Unfortunately it is possible that the classification as a whole is evaluated as good, although no cluster is recognized correctly.

Some validity measures use the fuzziness of the membership degrees. They are based on the idea that a good solution of a fuzzy clustering algorithm is characterized by a low uncertainty with respect to the classification. Hence the algorithms based on these measures search for a partition which minimizes the classification uncertainty. For example this is done using the *partition coefficient* [1].

Other validity measures are more related to the geometry of the data set. For example the *fuzzy hypervolume* is based on the size of the clusters [5]. Because in probabilistic clustering each datum is assigned to a cluster, a low value of this measure indicates small clusters which just enclose the data.

For fuzzy shell clustering algorithms other validity measures are used. For example the *fuzzy shell thickness* measures the distance between the data and the corresponding clusters [10].

4.2   Local validity measures

In contrast to global validity measures, local validity measures evaluate each cluster separately. Therefore it is possible to detect some good clusters even if the classification as a whole is bad.

An unsupervised fuzzy clustering algorithm based on local validity measures starts with a number of clusters greater than the expected number of clusters. A good recommendation is to use twice as many clusters as expected [10]. After each execution of the used fuzzy clustering algorithm the number of clusters and the data is reduced. Good clusters and the assigned data are temporarily removed. Bad clusters and data that represent noise are deleted. In a further iteration the fuzzy clustering algorithm is executed on the remaining data using the number of the remaining clusters. This is repeated until no cluster is left or no cluster is removed. After the number of clusters is determined, the temporarily removed clusters are fine tuned by running the clustering algorithm again.

Some local validity measures are derived from global validity measures. For example the fuzzy shell thickness can also be determined for each cluster separately [10]. However, sometimes the evaluation of a clusters varies depending on its size and completeness. For example it is difficult to distinguish between sparse clusters and cluster segments using these validity measures.

For two dimensional data good results are obtained by using the *surface density* [10]. The surface density measures the relation between the number of data assigned to a cluster and the number of data, if that cluster would be perfect. For example, for a full circle the number of assigned data is related to the circumference. It is also possible to distinguish between cluster segments and sparse clusters using the surface density.

For applications in computer vision it has been proven successful to use the surface density in combination with other local validity measures.

4.3 Initialization

The performance and quality of a classification computed by a fuzzy clustering algorithm depends to a high degree on the initialization of the fuzzy clustering algorithm. Especially fuzzy shell clustering algorithms are very sensitive concerning the initialization, because they tend to get stuck in local minima.

A widely used procedure for initialization is to start with some iterations of the FCM and, if applicable, the GK and the FCES. For instance, a good initialization of the FCQS and its modifications can be achieved by using 10 iterations of the FCM, 10 iterations of the GK and 5 iterations of the FCES [10]. In case only ellipsoidal shell clusters are searched, better results are achieved by omitting the GK because it sometimes tends to search for lines.

An alternative approach is to apply methods from computer vision. For example, some techniques from boundary detection can be used to initialize fuzzy shell clustering algorithms. One such algorithm is introduced in [7]. For shell cluster analysis it is superior to the initialization procedure using the FCM, the GK, and the FCES described above. The reason is that the initialization relates more to the nature of the searched clusters. In addition this algorithm estimates the expected number of clusters at the beginning of the fuzzy shell clustering algorithm. Hence it can reduce the computation time of the unsupervised fuzzy shell clustering algorithm. The results presented in section 5 are computed using this algorithm.

5 Fuzzy Shell Cluster Analysis in Computer Vision

Fuzzy clustering techniques can be applied in numerous fields. One of them is computer vision, in which clustering methods have been used for region segmentation for years. Another application in computer vision is contour detection and fitting or surface detection and fitting. Unsupervised fuzzy shell clustering algorithms seem to be well suited for this task.

The detection and recognition of boundaries in two dimensional pictures or surfaces in three dimensional scenes is one of the major problems in computer vision. A common method is the use of the generalized Hough transform which is able to deal with noisy and sparse boundaries or surfaces. However the disadvantage of the generalized Hough transform are its computational complexity and its high memory requirements[1] if there are only few assumptions concerning the boundaries or surfaces searched for. An alternative approach is to use fuzzy shell clustering algorithms, which perform boundary detection and fitting or surface detection and fitting simultaneously. These algorithms require far less computations and memory compared with the generalized Hough transform. Besides this algorithms are insensitive to local aberrations and deviations in shape as well as to noise.

Fuzzy shell clustering algorithms searching for quadrics or hyperquadrics can detect a large variety of curves, which they are able to detect. For many applications this is sufficient. It is recommended to use an unsupervised version of the FCQS that is based on the local validity criteria of the surface density [10]. However, the validity measure of surface density has to be slightly modified, because in applications of computer vision the aspect of digitization of images must be considered. That can be done using a correction factor [10].

Finally we present some results of boundary detection and recognition obtained by an unsupervised fuzzy shell clustering algorithm. Fig. 3 and 4 are obtained from fig. 1 and 2 respectively, by using an edge detection algorithm and an algorithm for line thinning. The data shown in these figures are divided into clusters by using an unsupervised fuzzy shell clustering algorithm. Fig. 5 and 6 show the clusters obtained by the FCQS.

It is obvious that the significant boundaries are determined correctly. Besides, like the CCM an unsupervised fuzzy shell clustering algorithm is able to distinguish

---

[1]The computational complexity is $O(n \times N_{p_1} \times N_{p_2} \ldots \times N_{p_{s-1}})$ and the memory requirement is $O(n \times N_{p_1} \times N_{p_2} \ldots \times N_{p_{s-1}})$, where n is the number of points, N_{p_i} is the number of quantization levels of the i-th parameter, and s is the total number of parameters [10].

between significant and insignificant structures. However, it is always important to bear in mind, that the computed classification is based on the choice of the validity and the distance measures. Therefore an optimal solution computed by an unsupervised fuzzy clustering algorithm sometimes differs from a classification obtained by a human. For example a human might describe the right border of the cup in fig. 4 using an extra line which is missing in fig. 6.

Summarizing unsupervised fuzzy shell cluster analysis is an interesting method for line detection and recognition or contour detection and recognition. Its advantage compared to other algorithms, e.g. the generalized Hough transform, is its lower computational complexity and its ability to distinguish between significant and insignificant structures. However, this ability is also a disadvantage because they cannot detect small and fine structures. Finally it is to remark that an optimal solution cannot be guaranteed.

Figure 1: picture of a disk

Figure 2: picture of a cup

Figure 3: contour of a disk

Figure 4: contour of a cup

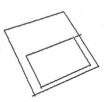

Figure 5: prototypes found by
the FCQS

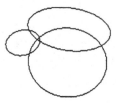

Figure 6: prototypes found by
the FCQS

References

[1] Bezdek, J.C.: Pattern Recognition with Fuzzy Objective Function Algorithms, Plenum, New York 1981.

[2] Bock, H.H.: Classification and Clustering: Problems for the Future, in: New Approaches in Classification and Data Analysis (Ed. Diday, E., Lechevallier, Y., Schrader, M., Bertrand, P. and Burtschy, B.), Springer, Berlin, 1994, 3-24.

[3] Davé, R.N.: Use of the Adaptive Fuzzy Clustering Algorithm to Detect Lines in Digital Images, Proc. Intelligent Robots and Computer Vision VIII, 1192 (1989), 600-611.

[4] Frigui, H. and Krishnapuram, R.: A Comparison of Fuzzy Shell-Clustering Methods for the Detection of Ellipses, IEEE Transactions on Fuzzy Systems, 4 (1996), 193-199.

[5] Gath, I. and Geva, A. B.: Unsupervised Optimal Fuzzy Clustering, IEEE Transactions on Pattern Analysis and Machine Intelligence, 11 (1989), 773-781.

[6] Gustafson, E.E. and Kessel, W.C.: Fuzzy Clustering with a Fuzzy Covariance Matrix, IEEE CDC, San Diego, Californien, 1979, 761-766.

[7] Höppner, F., Klawonn, F. and Kruse, R.: Fuzzy-Clusteranalyse. Verfahren für die Bilderkennung, Klassifikation und Datenanalyse, Vieweg, Braunschweig 1996.

[8] Krishnapuram, R. and Freg, C.P.: Fitting an Unknown Number of Lines and Planes to Image Data through Compatible Cluster Merging, Pattern Recognition, 25 (1992), 385-400.

[9] Krishnapuram, R., Frigui, H. and Nasraoui, O.: The Fuzzy C Quadric Shell clustering algorithm and the detection of second-degree curves, Pattern Recognition Letters 14 (1993), 545-552.

[10] Krishnapuram, R., Frigui, H. and Nasraoui, O.: Fuzzy and Possibilistic Shell Clustering Algorithms and Their Application to Boundary Detection and Surface Approximation — Part 1 & 2, IEEE Transactions on Fuzzy Systems, 3 (1995), 29-60.

[11] Krishnapuram, R. and Keller, J.: A Possibilistic Approach to Clustering, IEEE Transactions on Fuzzy Systems, 1 (1993), pp. 98-110.

[12] Krishnapuram, R. and Keller, J.: Fuzzy and Possibilistic Clustering Methods for Computer Vision, Neural Fuzzy Sytems 12 (1994), 133-159.

[13] Moore, J.J.: The Levenberg-Marquardt Algorithm: Implementation and Theory, in: Numerical Analysis (Ed. Watson, G.A.), Springer, Berlin, 1977, 105-116.

AUTOMATIC CONSTRUCTION OF DECISION TREES AND NEURAL NETS FOR CLASSIFICATION USING STATISTICAL CONSIDERATIONS

F. Wysotzki
Technical University of Berlin, Berlin, Germany

W. Müller and B. Schulmeister
Fraunhofer-Gesellschaft, Berlin, Germany

ABSTRACT

Two algorithms for supervised learning of classifications are discussed from the point of view of the usefulness of including statistical methods. It will be demonstrated that statistical considerations of very general nature (i.e. without assumptions on class distributions) can lead to substantial improvements of the learning procedure and the constructed classifiers. The decision tree learner CAL5 converts real-valued attributes into discrete-valued ones the number of which is not restricted to two. Pruning occurs during tree construction. The hybrid (statistical/neural) algorithm DIPOL solves the problem of choosing the initial architecture and initial weights by statistical methods and replaces additional hidden layers by a Boolean decision function. Both algorithms are also discussed within the framework of the ESPRIT-Project StatLog where about 20 of the most important procedures for classification learning are compared using statistical criteria.

INTRODUCTION

In this paper two algorithms for supervised learning of classifications are discussed mainly from the point of view of the usefulness of including statistical methods. It is generally argued that learning methods developed in AI and Artificial Neural Nets have advantages as compared with methods of statistical classification since they are "distribution free", i.e. no assumptions on the probability distributions of the classes have to be made a priori. In the following it will be demonstrated that using statistical considerations of very general nature (i.e. without assumptions on class distributions) can lead to several kinds of improvements of methods for supervised learning of classifications. In chapter 2 the algorithm CAL5 [UW81, MW94, MST94, MW96] for learning decision trees is briefly described which automatically converts real-valued attributes into discrete-valued ones. Using statistical considerations it is possible to construct attributes with an (in principle) arbitrary number of discrete values, locally restricted by the cardinality of the available training set only. Pruning occurs during tree construction and no postpruning is necessary. In the limiting case of very large training sets a uniform error bound for all decision nodes (leaves) of the tree can be defined on a given confidence level.

In chapter 3 the hybrid learning algorithm DIPOL [MTS94, SW96] is described which builds a piecewise linear classifier by a segmentation of the feature space into class or subclass decision regions in an initial step using classical statistical methods and optimizing the discriminating hyperplanes by learning in a second step. This way a well known problem of Artificial Neural Nets, i.e. the definition of the neural architecture a priori and initialization of weights is solved in a natural manner. Only one teachable layer is needed and additional hidden layers are replaced by a Boolean decision function (in combination with a maximum detector) constituting the output layer. CAL5 and DIPOL were included in the ESPRIT-Project Statlog [MST94] the aim of which was the comparison of about 20 of the most important algorithms for classification learning using about 20 data sets from different application areas. The performance of both algorithms will be briefly discussed within the framework of Statlog.

2. DECISION TREE LEARNING BY CAL5

The algorithm CAL5 [UW81, MW94, MST94, MW96] for learning decision trees for classification and prediction converts real-valued attributes into intervals using statistical considerations. The intervals (corresponding to discrete or "linguistic" values) are automatically constructed and adapted to establish an optimal discrimination of the classes in the feature space. The trees are constructed top-down in the usual manner by stepwise branching with new attributes to improve the discrimination of classes. An interval on a new dimension (corresponding to the next test refining the tree on a given path) is formed if the hypothesis "one class dominates in the interval" or the alternative hypothesis "no class dominates" can be decided on a user defined confidence level by means of the estimated conditional class probabilities. "Dominates" means that the class probability exceeds some threshold given by the user.

In more detail: If during tree growing a (preliminary) terminal node representing an interval in which no class dominates is reached, it will be refined using the next real-valued attribute x for branching. Firstly, all values of training objects reaching that terminal node are ordered along the new dimension x. Secondly, values are collected one by one from left to right forming intervals tentatively. Now by the user-given confidence level $1 - \alpha$ for estimating the class probabilities in each current tentative interval I on x a confidence interval $[\delta_1, \delta_2]$ for each class probability depending on the relative frequency n_c/n of class c and the total number of values n in the current interval is defined. A Bernoulli distribution is assumed for the occurrence of class symbols in I, i.e. a source delivering class symbols independently. Then for each tentative x-interval I the following „metadecision" is made

1. If for a class c

   $$S < \delta_1(n_c/n, n, \alpha)$$

   then decide „c dominates in I". S is a user-given threshold (for example S = 0.9) defining the maximal admissible error in class detection. If c dominates in I the path will be closed ("pruned") and c attached as a class label to the newly created terminal node.

2. If for all classes c in I
   $\delta_2(n_c/n, n, \alpha) < S$,

i.e. no class dominates, the tree is refined using the next attribute.

3. If neither 1) nor 2) holds I is extended by the next value of the order on x and the procedure repeated recursively until all intervals (i.e. discrete values) of x are constructed.

A special heuristics is to be applied in the case that (due to the finite training set) there are "rest intervals" on dimension x fullfilling neither hypothesis 1) nor hypothesis 2). The order of attributes for tree building is computed using the transinformation (entropy reduction) measure. The confidence interval $[\delta_1, \delta_2]$ is approximated using the Chebychev-Inequality, for computational reasons. (This results in somewhat too large intervals for each dimension x and in a loss of class discrimination. A finer approximation of the class boundaries -here by piecewise axis-parallel hyperplanes- may be obtained by storing tables of confidence intervals for the Bernoulli distribution and using table look-up. This leads to an increasing time for the learning phase).

Finally adjacent intervals with the same class label are joined. Note that in the case of very large training sets 1-S is a uniform error bound (on confidence level 1-α) for all terminal nodes of the final tree where hypothesis 1) could be confirmed.

If costs for not recognizing single classes are given, CAL5 uses class dependent thresholds S_c (c being the class not correctly recognized). From decision theory it follows [UW81] that one has to choose

$$S_c \approx const/cost_c,$$

where $cost_c$ is the cost for misclassification of an object belonging to class c to be given by the user. For details see [MW94, MW96].

By CAL5 the trees are automatically pruned during learning. Another difference to CART and the ID3-family (see [MTS94] for an overview and more detailed references) is that not only binary discrete valued attributes can be constructed by the learning algorithm but in principle (depending on the training set) attributes with an arbitrary number of discrete values (i.e. intervals). CAL5 was included in the ESPRIT-Project StatLog [MTS94] and performed very well on the data sets mostly suited for decision tree learning. In fact it was the best performing algorithm for the credit data sets.

Algorithm	Data sets not handled	Average rank on all data sets	Algorithm	Data sets not handled	Average rank on all data sets
dipol (s,n)		6.09	knn (s)		11.09
cart (s,r)	7	8.40	newid (r)		11.23
bay tree (s)		8.41	rbf (n)		11.45
cascade (n)	15	8.57	c4.5 (r)	1	11.48
logdiscr (s)		8.59	quadisc (s)		11.68
alloc80 (s)	1	9.41	cn2		11.73
backprop (n)	2	9.50	lvq (n)	1	12.29
smart (s)		9.82	ac2		12.73
indcart (r)		9.91	castle (r,s)	1	14.33
discrim (s)		10.23	naive bay (s)		16.23
cal5 (r)		10.82	itrule (r)	2	17.35
			kohonen (n)	3	17.42

Table 1. Ranking of StatLog-algorithms averaged on all data sets. s: statistical r: rule learning n: neural

Table 1 shows the average ranking of all StatLog-algorithms taken over all data sets. It was computed from the tables in [MTS94] where the results of StatLog for specified subsets of data sets are shown only. The basis of evaluation of the performance of an algorithm on a data set was the mean error using cross validation or mean minimal costs if cost matrices for not recognizing single classes were given. Note that CAL5 performs (on the average) slightly better than the ID3-extensions C4.5 and NEWID.

As compared with the other decision tree algorithms used in StatLog, the trees developed by CAL5 had - on the average - the least number of decision nodes, i.e. least complexity. In connection with the credit data sets it was discovered that it is expecially appropriate for tasks where it is optimal to "simulate" human decision making.

3. LEARNING A PIECEWISE LINEAR CLASSIFIER BY DIPOL

Another algorithm for classification learning having a statistical component is the hybrid (statistical/neural) algorithm DIPOL, which was also included in StatLog. It may be considered a kind of generalisation of CAL5 in the sense that it also constructs a segmentation (partition) of the feature space into decision regions with piecewise linear boundaries but using general hyperplanes and not axis-parallel ones only. DIPOL [MTS94, SW96] was developed within the project WISCON[1] for learning process control. A system following similar general principles was firstly described in [UW81].

DIPOL is a hybrid algorithm for classification learning which can be regarded as both

- a nonparametric and distribution free statistical approach with the special option of clustering classes or subclasses, respectively, optimizing the discriminating hyperplanes by gradient descent or stochastic approximation and carrying out the classification task on a symbolic level or
- a neural network approach with one teachable layer only, i.e. an extension of the PERCEPTRON- and MADALINE-principle [W62, N65] for piecewise linear discrimination by a statistically justified choice of the number and initial weights of neural units and fixed Boolean weights in a second (symbolic) decision layer.

In the first step by regression analysis (optionally combined with cluster analysis in the case of multimodal class distributions) discriminating hyperplanes for each pair of classes or subclasses, respectively, were constructed [SW96]. This leads to a segmentation of the feature space, where each segment (decision region for a class or subclass, respectively) can be described by Boolean variables, i.e. the signs of the normals of hyperplanes.

[1] Supported by the German Ministry of Science and Technology

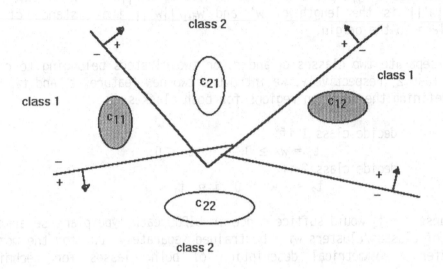

Fig 1: Discriminating Several Classes/Subclasses Pairwise by a Set of
       Hyperplanes (DIPOL)

Fig. 1 shows the representation of a generalized XOR. Thus the problem of
choosing the architecture and initial weights of artificial neurons is
solved in a way adequate for the problem at hand. The hyperplanes are
then optimized by batch- or one line-learning using an objective
function. The learning procedure is described in more details in what
follows.

LEARNING IN DIPOL

Let be $x = (x_1, \ldots, x_n, 1)$ a pattern vector and $w = (w_1, \ldots, w_n, w_{n+1})$ a
weight vector, then the scalar product

$$t = \sum_{i=1}^{n+1} w_i x_i =_{def.} wx$$

defines a new feature which is a linear combination of the original x_i,
$i=1, \ldots, n$, w_{n+1} is the threshold or bias, $x_{n+1} =_{def.} 1$ to simplify the
notation of the scalar product.

$t = 0$ defines a hyperplane with normal vector $w'/||w'||$, $w' = (w_1, \ldots, w_n)$, $||w'||$ is the length of w' and $w_{n+1}/||w'||$ the distance of the hyperplane to the origin.

Now to separate two classes c_1 and c_2 or two clusters belonging to class 1 and class 2, respectively, we introduce two new features t_1 and t_2, $t_2 = - t_1$, defining the decision regions for both classes:

decide class 1 if
$$t_1 = wx \geq 0 \quad \text{i.e. } t_2 \leq 0,$$
decide class 2 if
$$t_2 = - wx > 0 \quad \text{i.e. } t_1 < 0$$

(Of course $t = t_1$ would suffice since in DIPOL each hyperplane separating a pair of classes/clusters will be trained separately. But for the moment we prefer a symmetrical description of both classes for technical reasons).

An error for a pattern vector of class 1 occurs if

$$t_1 < 0 \text{ and } x \in c_1$$

and an error of class 2 if

$$t_2 = - t_1 \leq 0 \text{ and } x \in c_2.$$

DIPOL uses the objective function (error function)

$$F = 1/2 \left(\sum_{\substack{x \in c_1 \\ t_1 < 0}} t_1^2/||x'||^2 + \sum_{\substack{x \in c_2 \\ t_2 \leq 0}} t_2^2/||x'||^2 \right).$$

with $x' = (x_1, \ldots, x_n)$ for supervised learning.

F is computed by summing up on all misclassified patterns x of the training set. Note that the "error" $|t_{1,2}(x)|$ is - up to a scaling factor $||w'||$ - the distance of a misclassified pattern x from the discriminating hyperplane and that all hyperplanes relevant for classifying x are trained separately. For a more detailed discussion of DIPOL's error function see [SW96].

Using F the learning rule for batch learning (or "learning by epoch", i.e. using the whole training set simultaneously) is then gradient descent

$$w^{(l+1)} = w^{(l)} - \rho_1 \cdot \nabla_w F$$

$$\nabla_w F = \sum_{\substack{x \in C_1 \\ t_1 \in 0}} t_1 \cdot (x',1)^T / ||x'||^2 + \sum_{\substack{x \in C \\ t_2 \leq 0}} t_2 \cdot (x',1)^T / ||x'||^2$$

Incremental learning by stochastic approximation is defined accordingly using the single error only. Note that well known necessary conditions for the step size parameters ρ_1 must hold to establish convergence with probability 1. In general the ρ_1, $1 = 0,1 \ldots$ depend on the problem at hand,

$$\rho_1 = 1/1 \quad , \quad \lim_{1 \to \infty} \rho_1 = 0$$

may be a useful compromise. If costs for misclassifications of different classes are given by the user they can be explicitly included in F by multiplying the first sum in F by $cost(c_1, c_2)$ (cost for not recognizing class 1) and the second sum by $cost(c_2, c_1)$ (cost for not recognizing class 2), respectively.

An objective function linear in the new features $t_{1,2}$ is used in [UW81]:

$$F' = \sum_{\substack{x \in C_1 \\ t_1 < \varepsilon}} (t_1 - \varepsilon) + \sum_{\substack{x \in C_2 \\ t_2 < \varepsilon}} (t_2 - \varepsilon),$$

eventually with cost factors as above. ε is introduced because F' would be nondifferentiable at the hyperplane $t_1 = t_2 = 0$. This can be compensated by using the so called generalized gradient within the interval $[-\varepsilon / ||w'||, \varepsilon / ||w'||]$ around the plane (for details see [UW81]), where ε is a given parameter. This means practically that a fraction of correct classified patterns falling into the $[-\varepsilon / ||w'||, \varepsilon / ||w'||]$-band around the plane must be used for learning, too [We96].

For a single learning step stochastic approximation (incremental learning) is to be applied and one has to use the single step error

$$f' = \begin{cases} 0 & x \in C_1 \text{ and } t_1 \geq \varepsilon \text{ and } x \in C_2 \text{ and } t_2 \geq \varepsilon \\ t_1 - \varepsilon & x \in C_1 \text{ and } t_1 < \varepsilon \\ t_2 - \varepsilon & x \in C_2 \text{ and } t_2 < \varepsilon \end{cases}$$

The learning rule is given by

$$w^{(l+1)} = w^{(l)} + \rho_1 \nabla_w f'$$

as above (note that F', f' are negative, hence the plus sign) with

$$\nabla_w \cdot f' = \begin{cases} 0 & x \in C_1 \text{ and } t_1 \geq \varepsilon \text{ or } x \in C_2 \text{ and } t_2 \geq \varepsilon \\ x' & x \in C_1 \text{ and } t_1 < \varepsilon \\ -x' & x \in C_2 \text{ and } t_2 < \varepsilon \end{cases}$$

and

$$\frac{\partial f'}{\partial w^{n+1}} = \begin{cases} 0 & \text{as above} \\ 1 & x \in C_1 \text{ and } t_1 < \varepsilon \\ -1 & x \in C_2 \text{ and } t_2 < \varepsilon, \end{cases}$$

$x = (x_1, \ldots, x_n, 1) = (x', 1)$, $w = (w_1, \ldots, w_n, w_{n+1}) = (w', w_{n+1})$. It is interesting to note [UW81] that in the limiting case $\varepsilon = 0$ and $\rho_1 = 1$, $l = 0, 1, \ldots$, this rule gives the classical PERCEPTRON learning algorithm for separable classes (see for example [MP69]):

If the error is caused by a too small $t_1^{(1)} = w^{(1)}x$ then $w^{(l+1)} = w^{(1)} + x$ (i.e. $t^{(l+1)} = t^{(1)} + ||x||^2$) otherwise if $t_2^{(1)} = -t_1^{(1)}$ is too small (i.e. $t_1^{(1)}$ too large) then $w^{(l+1)} = w^{(1)} - x$.

We performed experiments [We96] by replacing in DIPOL F by F' and f by f', respectively, for some of the Statlog-data sets using the Statlog evaluation procedure [MST94]. There were no essential differences between the results for both types of objective functions F and F', respectively. But the learning time increases in the second case since the ε-fraction of correct classified patterns has to be computed and used for adapting the weight vector.

Incremental learning by DIPOL can also be applied to the case of a dynamically changing environment, i.e. if the class distributions change in time [We96]. However the change must be slow as compared with the learning speed and must not affect the topology of the segmentation of the feature space defined by the set of hyperplanes. Note that DIPOL shows the best average performance on the StatLog-datasets (Table 1), it was also the neural net algorithm with the highest learning speed [MST94].

DECISION OF CLASS MEMBERSHIP IN DIPOL

Let be x a pattern vector (not necessarily belonging to the training set) to be classified by means of the set of hyperplanes constructed by DIPOL. If there are m classes or subclasses (clusters), respectively, then N = m(m - 1)/2 hyperplanes are needed. In the following the hyperplanes are enumerated from 1, ..., N and we use $t^{(k)} = t_1^{(k)}$ to define the k-th hyperplane fixing the direction of its normal arbitrarily. We define a N-dimensional vector V(c) of three valued attributes depending on a class/subclass variable c by

$$V_1(c) = \begin{cases} 1 & \text{if } c = c_i \\ -1 & \text{if } c = c_j \\ 0 & \text{otherwise} \end{cases}$$

I.e. the normal of the 1-th hyperplane (1 = 1, ..., N) points to the decision region for class c_i if hyperplane 1 is relevant for discriminating the pair of classes (c_i, c_j) (or clusters belonging to different classes, respectively), and $V_1 = 0$ if it is irrelevant for discriminating (c_i, c_j). That means the set of vectors V(c), c ∈ $\{c_1,$ ...., $c_m\}$ is equivalent to a Boolean function (for example in disjunctive normal form) describing the "topology" of the segmentation of the feature space with don't care variables if $V_1 = 0$. For example in Fig. 1 we may define by enumerating the hyperplanes anticlockwise and starting with that one discriminating c_{12} and c_{22}

$$V(c_{12}) = (\ 1,\ -1,\ 0,\ 0)$$
$$V(c_{21}) = (\ 0,\ 1,\ 1,\ 0)$$
$$V(c_{11}) = (\ 0,\ 0,\ -1,\ -1)$$
$$V(c_{22}) = (-1,\ 0,\ 0,\ 1)$$

This corresponds to the Boolean functions

$$V_1(c_{12}) \wedge \sim V_2(c_{12}) \vee \sim V_3(c_{11}) \wedge \sim V_4(c_{11}) \rightarrow \text{class 1}$$
$$V_2(c_{21}) \wedge V_3(c_{21}) \vee \sim V_1(c_{22}) \wedge V_4(c_{22}) \rightarrow \text{class 2}$$

In DIPOL for a pattern $x = (x_1, \ldots, x_n, 1)$ an attribute vector

$$st(x) = (st_1(x), \ldots, st_N(x))$$

with

$$st_k(x) = sign(t^{(k)}) = \begin{cases} 1 & t^{(k)} \geq 0 \\ -1 & t^{(k)} < 0 \end{cases}$$

is used for classification in the following manner: Decide for that class $c = c_j$ which maximizes the scalar product

$$S(x,c) = V(c) \cdot st(x) = \sum_{k=1}^{N} V_k(c) \cdot st_k(x)$$

or for which $S(x,c_{jr})$ is maximum for some r, respectively. The latter case occurs if c_j has a multimodal distribution, i.e. consists of several subclasses found by the optional cluster analysis in the initial step. (For details of the cluster algorithm for which any standard procedure can be used, see [SW96, UW81]).

There may be segments with no class assigned because there are no training patterns falling into them (see for example the "empty" segment in the middle of Fig. 1). In this case the class of x is decided using the minimum distance to segments for which classes are defined. (The description of DIPOL's decision procedure given here is somewhat oversimplified, for more details see [SW96]).

Note that for the decision procedure N "linear output neurons" computing a scalar product with fixed weights $\in \{-1, 0, 1\}$ and at least an additional maximum detector are needed. This (not teachable) output layer replaces the single output neuron of the MADALINE-system [W62, N65] which is able to compute a Boolean majority decision function only. Thus MADALINE cannot learn some relatively simple classifications [UW 81]. It also replaces additional teachable hidden layers used for example in Backpropogation.

4. CONCLUSIONS

Two algorithms for supervised learning of classifications have been described which integrate methods developed in Artificial Intelligence and Artificial Neural Networks, respectively, with statistical ones. CAL5 learns decision trees by transforming real-valued attributes into discrete-valued ones (intervals) the number of which depends on the conditional probability distributions of classes as well as on the available statistics but is not restricted apriori. Whereas CAL5 (like other decision tree algorithms) constructs decision regions in the feature space using axis parallel hyperplanes only, the piecewise linear classifiers built by the hybrid learning algorithm DIPOL can be considered as some kind of generalization of the segmentation delivered by CAL5 in the sense that hyperplanes in general position are used. Since the "topology" of segmentation into class/subclass regions constructed by DIPOL in the initial step (based on statistical methods) can be described by Boolean functions and is maintained during the optimization of the hyperplanes in the learning phase, class decisions can be made by means of logical (Boolean) variables, the meaning of which is defined by the signs of normals of the hyperplanes. Since DIPOL uses one teachable layer of artificial neurons only this leads - at least in principle - to the possibility of extraction of interpretable rules in special cases which may be useful for Data Mining. Both algorithms were included in the ESPRIT-project StatLog and performed - on the average - very well on the data sets used there.

REFERENCES

1. [MST94]   Michie, D., Spiegelhalter, D.J. and Taylor, C.C. (Eds.): Machine Learning, Neural and Statistical Classification, Ellis Horwood, New York 1994.

2. [MP69]   Minsky, M. and Papert, S.: Perceptrons, MIT Press, Cambridge, MA, 1969.

3. [MW94]   Müller, W., and Wysotzki, F.: Automatic Construction of Decision Trees for Classification, Annals of Operations Research, 52 (1994), pp. 231-247.

4. [MW96]   Müller, W., Wysotzki, F.: The Decision Tree Algorithm CAL5
             Based on a Statistical Approach to its Splitting Algorithm.
             in: Nakhaeizadeh, G. and Taylor, C.C. (Eds.), Machine
             Learning and Statistics, The Interface. John Wiley & Sons,
             New York, 1996.

5. [N65]   Nilsson, N.: Learning Machines. McGraw-Hill, New York 1965.

6. [SW96]   Schulmeister, B. and Wysotzki, F.: DIPOL - A Hybrid Piecewise
             Linear Classifier, in: Nakhaeizadeh, G. and Taylor, C.C.
             (Eds.), Machine Learning and Statistics, The Interface. John
             Wiley & Sons, New York 1996.

7. [UW81]   Unger, S. and Wysotzki, F.: Lernfähige Klassifizierungs-
             systeme, Akademieverlag, Berlin 1981.

8. [W62]   Widrow, B.: Generalization and Information Storage in
             Networks of ADALINE "Neurons", in: Self Organizing Systems
             (M. C. Yovits, G. T. Jacoby, and G. D. Goldstein, eds),
             Spartan Books, Washington, D.C. 1962, pp. 435-461.

9. [We96]   Werk, R.: Untersuchungen zur Korrektur und zum inkrementellen
             Lernen von Klassifikationen durch den hybriden Algorithmus
             DIPOL, Diplomarbeit, Dept. of Computer Science, Technical
             University of Berlin 1996.

FROM THE ART OF KDD TO THE SCIENCE OF KDD

Y. Kodratoff
University of Paris-Sud, Orsay, France

ABSTRACT

It has been already largely proven that **Knowledge Discovery in Databases** (KDD) is an interesting new research field, able to provide financial returns to the companies that are willing to invest into it. This fact demonstrates the excellent social value of KDD. A Science, however, is not uniquely defined by this feature. It needs also to show an internal logic, due to a specific approach to the real-life problems it deals with. This last point of view has been less emphasized in the existing KDD literature. This paper attempts, without any pretense to be exhaustive, to start filling up this gap. We shall explain why KDD is not just "a bunch of techniques" but a real Science, certainly one still under organization, but which shows the strong inner motivation that other Sciences do. In conclusion we shall give a compact definition of KDD, and show what is the concept it provides measurement of, as a function of which other concepts.

1. INTRODUCTION

1.1 Controversial KDD?

Each time a new set of techniques develops, the same pattern of behavior appears inside the scientific world. Scientists tend to divide in two groups. The conservative one claims that the old techniques were good enough to take care of all problems. The enthusiastic one jumps on the new techniques, develops them, starts formalizing them, and, when this stage is achieved, the conservatives begin appreciating the new technique, and acknowledge it as a new scientific field. There have been recently two very striking examples of this process in Statistics and Computer Science (CS). Statistics are nowadays becoming even somewhat doctrinal, while, still in the seventies, proper research work was done under the name of Probability Theory, and Statistics were just a "bunch of techniques". About CS, how often did I hear (and even read in a "serious" French review), until the end of the sixties: "the meeting of a soldering iron and of Logic does not make up a Science!" More controversial examples, at present, are Artificial Intelligence (AI) and Fuzzy Set Theory, still widely despised by many orthodox scientists. AI has been declared often a pre-science, as alchemy has been to Chemistry, and Fuzzy Set Theory is looked upon as useless, "since everything you do with it, you could do it with proper Statistics." Both slowly gather nevertheless scientific recognition by some of the more adventurous people of the conservative group, mainly under the pressure of their applications.

Since **Knowledge Discovery in Databases** (KDD), also called in industrial environments **Data Mining** (DM), is born barely some five years ago [1], it is not surprising that many scientists still claim that it is not a Science but "only" an engineering topic. This paper holds the position that, since KDD is still so young, indeed many of its features are still blurred, and it looks nowadays more like an Art than a Science. Nevertheless, it is a "art" as was "the art of computer programming" twenty years ago, i.e., under the process of generating a Science.

The aim of KDD is finding a convenient tradeoff between the amount of data and the amount of information provided to people that need to take a decision. For this, several existing techniques will be used, discovery of relationships inside data, **Pattern Recognition** (PR), **Data Analysis** (DA) and exploratory statistics, and **Machine Learning** (ML). The techniques have to be joined in an Expert System technology so as to automatize the discovery of knowledge (or information) buried into (typically) very large databases. The whole should show the comprehensibility required from modern induction systems [2].

We shall define what KDD is, and we shall analyze its several tasks, especially those that set up new problems to be solved by forthcoming scientific research, not simply by putting together existing techniques. When looking for this definition, we must keep in mind that the goal of all existing techniques, ML, Statistics, but also **Database Management Systems** (DBMS) and **Visualization Systems** is obviously to help the user in handling his data, thus all of them can claim that they have been trying to solve for ever the same problems as KDD does. They have been however developed by very different scientific communities, for some good reasons.

1.2 What KDD is and its three components

Let us follow Fayyad et al. [3] to define **Data Mining** (DM) as using any kind of algorithm in order to find patterns into the data. DM is thus one step of the more complex process of KDD, which aims at transforming the data into knowledge, through a process that begins by accessing the data, follows by mining it, and ends in interpreting it, usually with the help of a visualization tool. The algorithms used by DM are of two different kinds. They can be more "numeric" oriented, and they stem from **Statistics**. They can be more "symbolic" oriented, and they stem from ML. When the information to gather is of visual nature, then **PR** techniques are used. In this way, KDD makes use of DM without caring much about the origin of the pattern extractor, it is a KDD problem to choose the appropriate DM tools, and to make sense of the discovered patterns. Some DM techniques have been concerned by this problem in the past, they can be seen as precursors of KDD.

DBMS take into account very large amounts of data, and attempt to ease up the work of the user to query this data. Their main concern is an efficient management of the data in collecting, updating and retrieving it. The resulting languages, such as SQL queries, are rigorous and efficient, but tend to be hard to use. Helping the user to write down SQL queries is a part of DB research which is very relevant to KDD. It is to be expected that using a more powerful language, such as object oriented or deductive DB will largely modify the problems in KDD.

The main difference between KDD and DBMS is that the last are concerned with deductive queries only (what can I deduce from the data?) and not at all by inductive ones (what can I induce, or learn, or guess from the data?). In KDD, inductive queries must be allowed, without leading to an untractable obscure query language, or to lengthy answering time.

Another KDD problem follows from the fact that there exists some meta knowledge relative to the data, its model. Using the data model during mining seems to be a very difficult problem since so few works address it.
What is a usable DB?

Statisticians are also concerned with rigor and efficiency and they obtain strong results that are all biased by this concern. Their results are usually hard to understand (one says that DA produces more data than it started from, making it quite difficult for a non-specialist to understand it), and can be even misleading when a non statistician wishes to interpret naively the results of his statistical packages. Statisticians, as opposed to people in Databases (DB), are interested in deductive as well as in inductive reasoning (at least those performing exploratory statistics), and one can discover statistically significant previously unknown information from their results. Statistics have been based on exclusive handling of numeric data, which is no longer true nowadays. Still, we want to emphasize here that they have been thought and created with this bias.

A very positive point about statistical methods is that they are designed to deal with data showing the following typicality: they are large, noisy, time-dependent, redundant, with missing values, and heterogeneous (symbolic and numeric). Existing and well-known methods of classification and discrimination come from DA (or, more generally, from exploratory statistics). Here, we shall speak to some length of an important application of Bayesian statistics, the system AutoClass [4].

Nevertheless, these techniques need to improve in three directions. Firstly, knowledge extracted from data should be understandable directly by the domain expert. Secondly, data are not simple tuples in a table, but they are structured following the data model which should be used during the induction step. Thirdly, some domain knowledge can be found in commentaries added to the data. All this knowledge should be exploited during the discovery process, and this is still one of the challenges unsolved in the state-of-the-art of KDD research.

People in **ML** address a quite different problem. They are concerned almost exclusively by induction; they do use deduction, but only as a tool that can help the overall inductive process as happens in **Inductive Logic Programming** (ILP). They tend to take into account the concern for comprehensibility: a field expert should understand the results provided by a ML program with little effort. They are dealing usually with symbolic and numeric data, even though they historically begun by using pure symbolic data, as opposed to Statistics. *Symbolic* means here (numbers are indeed symbols!) that the data have semantics different from the one of numbers and that the features can take only a finite number of discrete values. This symbolic data shows relationships expressed by *part-of, is-a, cause-of, has-property,* etc. relations. For instance, one takes into account that the concept *dolphin* is a (is-a relationship) kind of *mammal*. This fact, and the inheritance on the properties of a mammal to a dolphin (for instance, one knows even without having met this behavior, that dolphin ladies milk their babies). Some other relations have to be expressed by theorems. For instance, expressing that almost all sons have a mother and a father (with the exception of a few Celtic heroes) establishes a non hierarchical relation between the concepts of *son*, *mother*, and *father* which would be very difficult to describe in a numeric way. These relationships and theorems are often called **domain knowledge,** or **background knowledge**, and part of ML efforts have been to include explicit background knowledge in the inductive process, while Statistics includes it implicitly. Yet another concern of ML is the one of bias. Induction is a very biased process, depending heavily on context, and making explicit the bias of each of the learning programs is an integral part of the ML research program.

ML is thus tuned to take into account heterogeneous data, and to show some comprehensibility. This last point has not been however the main concern of ML specialists, it still needs a large amount of research. Contrary to what is often claimed, ML does take into account noisy data. Inversely, it is quite true that many programs cannot handle large amounts of data because of the complex treatment needed to transform them into rules summarizing the data.

Computational efficiency and comprehensibility are two issues that need to be improved in ML programs so that they become more helpful in the KDD process.

1.3 Why is KDD born?

KDD is born when overwhelming amounts of data have been gathered, and that managers realized that no existing technique was able to help them interpret this data. For instance, one of the most famous KDD result is a "simple" adaptation of inductive techniques using information compression to a problem that was deemed helpless without computer aid: the interpretation of astronomical surveys. The adaptation of the ML technique to the problem might have been very clever, still it did not shake the bases of ML. On the contrary, the realization of a tool, encompassing all steps of interpretation (in passing, but not essential to our point: with a better precision than a human could achieve given as much time as he required), was more than a new tool, it was showing an essential progress, namely the understanding that a KDD program can interpret data starting from their raw form, up to a level directly understandable by a specialist of the field (i.e., what is called **knowledge** by astronomers), without introducing a computer scientist at any stage of the process.

Existing techniques, be them Statistics or ML, rely on a kind of cycle during which the DM specialist makes recommendations to the field specialist. In most cases, the cycle goes as follows: the field specialist gathers data, consults the DM specialist who recommends some changes in the way data is gathered, or recommends to gather new data, in order to insure that his DM tools work correctly. The new changed data are then submitted to the scrutiny of the DM tools, and their result is analyzed jointly by the DM and field specialists, in order to determine what, in the obtained results, comes from the data, what comes from biases in the data, and what comes from biases in the DM tool. Data and/or DM tool are then again revised in order to make these bias harmless to the field specialist.

This cycle has simply become impossible to perform for data that is gathered in a unique run, as astronomical data is, or in way that cannot be changed because of the cost of a change, as for most commercial data. The purpose of KDD is to get rid of this cycle by providing the user with a tool that will adapt itself to the data (instead of asking the data to adapt to the tool), maybe under some field specialist guidance, and at any rate without DM specialist help. This task is entirely new, asking for a new Science. In reality, this task has been partly performed in the past, especially when field specialists became themselves DM specialist, but it has never until now been looked upon as a specific scientific problem. In other words, we call precursors all these specialists that made in themselves the link between their field and DM: they solved parts of the problem, but they never posed it as a full-fledged scientific problem, that can be solved without asking humans to gather both types of knowledge.

1.4 Is KDD ungrounded?

There is a standing debate opposing scientists on the definition of knowledge. For many, knowledge cannot be separated from its bearer because formal knowledge (as the one found in books) is intricately merged to the experiences of the individual that carries this formal knowledge. They claim that knowledge is irreducible to formal knowledge because the latter is "ungrounded", i.e., unrelated to life experience. The opposite side claims that formal knowledge is enough to represent adequately and efficiently what is needed of human behavior and thinking. Illustration of this debate is found in a new approach to Turing's test [5], as presented in Searle's "Chinese room" [6, 7]. A computer can provide perfect translation of Chinese, without having real knowledge of Chinese, because this knowledge includes all facts of life a computer ignores.

This debate is still going on. In order to show how far it could lead us, we would like to cite here a philosopher of the sciences, M. Draganescu [8], who thinks that Science as it stands is able to deal with formal knowledge, but not with the fully grounded one, that he calls "conscious thinking". In order to build up a Science able to deal with all aspects of knowledge (he calls it: "structural-phenomenological"), he has to go as far as rethinking the origin of the Universe, starting it from an original "ripple of information", rather than starting, as astronomers do, from an explosion of energy. He thus presents a new cosmological vision necessary to cope with grounded knowledge.

We shall conclude here, somewhat cowardly, by saying that indeed Draganescu's structural-phenomenological Science does not exist yet, and we must do our best with Science as it stands, relying only on formal knowledge. In other words, we agree that formal knowledge does not exhaust knowledge, but we still want to go on, knowing that we are affected by one deep weakness. In the conclusion of this paper, our definition of knowledge, in KDD, will reflect this essential weakness.

1.5 Real-world knowledge discovery

One deals with knowledge discovery, one must thus explore the data (with a variable amount of help from the user) in order to find "interesting" patterns that help to express an information that was implicit in the data. The deep problems under the concept of knowledge discovery are the following:
- knowledge representation,
- attribute selection,
- using data that can be incomplete, noisy and badly distributed,
- defining what might mean "interesting", "useful" or even "surprising".

As it has been underlined by Brachman and Anand [9], this process is not just a crude exploitation but a complex process, nearer to archeology than to mining. Going on with the archeology analogy, one can say that knowledge discovery takes place in three steps. Making these steps more precise will help us to define what we mean by "real-world" KDD. Many claim that uncertainty, or missing data, or huge size of the DB, or else make their data "real-world". We disagree with these statements since one can find in academic repositories data of all kinds.

The first step, analogous to archeological site selection, is the one of data selection (also called windowing) in order to select the place best to work in the database. The work of Wirth and Reinartz [10] illustrates how much this step can be important in large real-world applications. Their work even shows that mining at the wrong places can be totally misleading since the causes of car breakdown vary with the profession of the car users. In a sense, finding the good contexts into which KDD will yield coherent results is an essential part of the application. Academic research ignores this problem, or calls it a trivial matter, while it is both difficult and interesting.

The second step is the necessary cleaning of the data that puts it under such a form that dataminers can use. In archeology, this step, however not striking, is very important since it underlines the important features, and blurs the non important ones. In KDD as well, this step is often overlooked. In most cases, however, it makes the difference between what is "academic" and what is "real-world". The deep difference is that in real-world, one knows that there is a phenomenon happening, that information about it is gathered in the DB, but one does not know exactly what are the factors to use (the input features of the problem are not all defined), and one does not know exactly what is to be induced (the conclusions of the field-experts are very complex, and not easy to model by a few output features). A good example of this problem is found in [11] who explains, for instance, that a production incident is a complex mixture of facts. Pointing at some of them as being the features of the problem is already a difficult problem, upon which relies the success of subsequent mining.

The third step corresponds to the interpretation of the scattered archeological remains into a coherent model. Some KDD systems are nothing but clever visualization systems, that ease up the interpretation of the user. They obviously do not treat the whole KDD problem, but their cleverness is itself a very hard KDD problem. A solution, classical in DA, however heavy, is to have several models, and to find the one that best fits the data. Two well-known instances of this behavior take place in regression and clustering. One allows several polynomial forms in regression, and several possible values for the number of clusters. Again, building possible models, finding the key features that differentiate among models, all that is also part of a real-world KDD application.

We see that the research program presented above has already been, at least partly, undergone by PR. This field shows also the concerns for comprehensibility, speed, etc. that are characteristic of KDD. The main difference between KDD and PR is that KDD detects logical patterns in the data, while PR looks for visual patterns. There are so many differences between the two problems that KDD and PR cannot be confused presently. It might turn out, however, that visual and logical patterns show deep and still unknown commonalties. This could result in the long range in a possible merge of PR and ML which show clearly the same opportunistic behavior in front of data.

1.6 First paradoxical feature of KDD

Let us recall the fact that KDD takes care of the whole process of transforming data into knowledge. The word *knowledge* itself is obviously badly defined unless a user can define what "makes sense" for him. For instance, useful knowledge is not defined in the same way for a technician and a manager. The KDD process is therefore user and goal oriented. The same user, working in different contexts to achieve different goals, may have different definitions of what is meaningful and useful to him. On the other hand, KDD is supposed to take care of huge amounts of data, thus making automation unavoidable. One touches here *the first paradoxical feature of KDD.* On the one hand, it is user-oriented, and the user must be included in the loop which extracts knowledge from the data. On the other hand, the amounts of data are so large that full automation is necessary. This contradiction contains the essence of many scientific problems of KDD whose role is to manage these contradictory needs.

1.7 Second paradoxical feature of KDD

The three components of KDD: DA, DBMS, and ML, show quite different concerns, as described in section 1.2 above. It is thus quite normal that these three scientific fields have been evolving in different directions. The *second paradoxical feature of KDD* is that it demands to use in a coordinated way those three scientific fields of various concerns. One cannot seriously argue that one has "just to merge" them. In particular, they evolved knowledge representations that catch different aspects of reality, and that are almost impossible to reconcile without losing the strength of the concerned field. Taking care of these divergent knowledge representations without losing too much efficiency is one of the major problems of KDD, and it has been almost ignored up to now. An exception is found in [11] and [12].

It follows that, from a technical point of view, KDD can be seen as standing in the center of a triangle formed by Statistics, ML, and DBMS as figure one below illustrates.

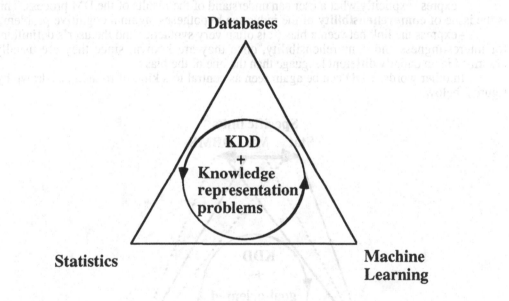

Figure 1. The second paradox of KDD. KDD must merge together DBMS, Statistics, and ML which evolved very different knowledge representation schemes, and which have very different concerns.

1.8 Third paradoxical feature of KDD

Statistics, ML and Databases are all three under the pressure of the technical demands, and all three have some links with Cognitive Science. However, it is clear that Databases are strongly user driven. Statistics are also widely used, while ML has found of the order of 100 applications over the world, which is not negligible, but very little as compared to the applications of its two companion domains. On the contrary, many ML specialists have been concerned by the relevance of Cognitive Science, even if this trend has not been enough developed in the past. KDD should put into existence links that are still very weak, and unify those links in such a way that, for instance, one does not start building specific links between Statistics and Cognitive Sciences, but between KDD and Cognitive sciences. This is what we call *third paradoxical feature of KDD* by which KDD is supposed to create and unify the links between existing technologies, the industrial demands, and human ways of reasoning. The main problem it meets in achieving this goal is the hard problem of **goal-oriented DM**. The aim of KDD is to include the user's goals and needs in a ML program, a statistical package, a DB query. This implies usually specific **bias** relative to human goals. These bias are very implicit in the DBMS and the statistical packages, somewhat more studied in ML. In none of these fields, however, are solved the three following problems:

- express explicitly the induction bias and measure their impact on the result of the learning systems. This is a problem for DM research.

- express explicitly the goals of the user. This is typically a cognitive problem since very often users can hardly express their real goals in words. One needs to analyze their actions to discover their real goals. This is the issue of **interestingness** of the inductive hypotheses, a topic never dealt with except in the KDD literature (see examples, *infra*).

- express explicitly what a user can understand of the results of the DM process. This is the issue of **comprehensibility** of the inductive hypotheses, again a cognitive problem.

- express the link between a bias (it is often very syntactic) and the user's definitions for interestingness and comprehensibility, once they are known, since they are usually described in an entirely different language than the one of the bias.

In other words, KDD can be again seen as central to a kind of triangle, as drawn by figure 2 below.

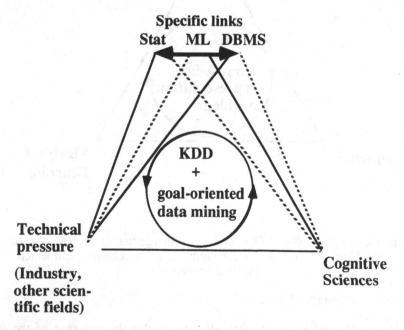

Figure 2. The third paradox of KDD. KDD is central to academic research, industry needs, and human understanding. This is linked to the problem of goal-oriented data mining.

Led by these three paradoxical features, and by the images of figures 1 and 2, we shall analyze in the next sections some of the tasks of KDD that are obviously opening a new research area, thus making of KDD a new Science.

It must be said that most DA specialists and statisticians are quite aware of the issues of interestingness and comprehensibility since they are working on real world applications. Nevertheless, the bulk of the research in their field is not oriented towards finding a solution to these issues, as KDD attempts to do. ML has known some more many attempts to deal with comprehensibility, such as explained in Michalski's "comprehensibility principle" [13], for instance, but one must confess that most of the field works on increasing precision as DA and Statistics.

Another question in debate is whether or not including the field of Knowledge Acquisition (KA) as a component of KDD. KA has been building very complex and useful models of human problem-solving activity. The purpose of the field is indeed some kind of comprehensibility, but it is oriented towards a better understanding of the knowledge gathered, so as to allow easy changes and adaptation to new problems. KDD wants to provide comprehensible results to its users, possibly with the help of programs the functioning of which can be quite obscure. This is opposed to KA that might yield very obscure results coming out of programs the functioning of which must be clearly understood. This is why we do not believe that the triangle of figure 1 should be enlarged to a square having KA as one of its angles.

2. A NEW POINT OF VIEW OF CLASSICAL ML PROBLEMS

As said above, it is clear that including induction in DBMS, and comprehensibility in Statistics is a very difficult task since these fields have not be developed with this aim. This will be the topic of section 3 and 4. Let us start by giving some indications of how the field of ML will be modified in order to fit into the KDD frame.

Since some 20 years that ML exists, it has developed its own way of looking at automatic knowledge acquisition. As a quite accurate approximation, one can say that among the many trails that have been tested by the ML scientific community, mainly two had such a success that they spread onto other fields. They are the so-called supervised and unsupervised learning paradigms. The novelty introduced by the KDD approach is mainly in putting emphasis on unsupervised learning (as opposed to most of ML published papers), and in stressing comprehensibility in supervised learning. A third ML theme, the discovery of numerical relationships in data has received much less attention. Its importance in KDD is such that we will have to say a few words about it at the end of this section.

Supervised learning amounts to the discovery of intentional description of classes (called also clusters, or also concepts), a partial extentional description of which is provided to the learning algorithm. In other words, the class is known and described by a few of its instances; the algorithm is supposed to build out of them a more general description that can be used in order to classify appropriately new instances of unknown class. For instance, suppose that the data are relative to cancer stricken patients, and that the classes are the set of possible treatments that have been successful in the past. The goal of the learning algorithm is to be able to decide, for a new patient, what will be the best appropriate treatment. For

this, one will have to find an intentional description of the past success cases. This goal can be reached in many ways, the most explored of which are trained neural networks (NN), decision trees, or rules such as
IF (set of the patient characters) THEN (appropriate treatment).

Unsupervised learning, also called classification by DA (an early reference is [14]), starts with an unstructured set of examples, and it provides classes (or clusters) out of these unstructured data. In this case, the class is unknown, and the goal of learning is to find significant clusters (the classes) of examples belonging to the same class, which is a partial extentional definition of the class.

To summary, unsupervised learning provides extentional descriptions of the classes, and supervised leaning provides intentional description of the classes. It is clear that the two paradigms can be composed, but the fact is that most authors have been concentrating on one or the other of them.

2.1 Unsupervised learning (AutoClass, COBWEB)

Two systems are now very much in use in the KDD world, even though their building is due to a small subclass of ML specialists, who are dealing much more often with supervised learning. One is AutoClass (a (strictly!) Bayesian classifier [4] and COBWEB [15]. Let us now see what are these two unsupervised learning algorithms.

COBWEB builds class hierarchies induced by a so-called *utility measure*. Given the present classification and a new example, COBWEB measures which of the existing classes should include the new example, or, alternately, if a new class should be created, by computing the utility of each of these actions. This is very similar to classical DA algorithms, with this difference that, in COBWEB, the distance measure is provided with the algorithm (it is the utility measure), and that the user does not need to specify how many classes are desired. The original implementation of COBWEB deals with discrete values only, and complete sets of data (no missing values). In the case of continuous and/or missing values , the user must discretize the continuous data (by consulting an expert), and/or approximate the missing values. There exists implementations that directly deal with these problems: CLASSIT [16] and COBWEB/3 [17]. These implementations hypothesize a normal distribution of continuous data . Such a continuous "COBWEB" has been successfully used on a large set of visual data [18, 19]. This clearly shows that COBWEB is not useful for small data sets only, even though it is indeed a slow algorithm: it is incremental, which allows to add constantly new examples in the hierarchy.

AutoClass is a classifier that makes hypotheses on data distribution, that computes its distance measure according to this distribution. As in COBWEB, there is no need to provide a distance measure, nor the desired number of classes (one must provide a maximal number, not the exact one). In practice, besides well-represented classes, AutoClass produces usually also a few badly-represented ones. An iterative use of AutoClass, decreasing progressively the maximum number of classes, allows thus to find the best number of them. AutoClass generates non hierarchical classes. It is however clear that, applying AutoClass recursively, one can also obtain a hierarchy of classes. The main two differences with COBWEB are that the user must provide the depth of the hierarchy (which is the number of recursive applications of AutoClass), and, more importantly, that the hierarchical classes are computed independently from each parent class. On the contrary,

adding a new class in COBWEB is performed by comparing all classes appearing in an existing tree.
In other words, when hierarchical classes are necessary, one should prefer COBWEB. Inversely AutoClass should be chosen for all other cases since it is much faster and, since it is theoretically better grounded, it provides better explanations of its results.

2.2 Learning class descriptions in intention

As we already recalled, it is a fact that ML as a whole has been working mostly on the topic of supervised learning. Its best results are found in C4.5 [20] for attribute-value data, and ILP for relational data. In this paper, we shall also describe SIAO [21], a genetic algorithm that generates relational formulas covering a set of examples. KBG [22, 23] could be as well used, but it is very slow, hard to parallelize (as opposed to SIAO), and it can work on small sets of examples only (say, at most some 100 examples).

2.2.1 Generalization by discrimination (CART, C4.5)

Both approaches, the statistical one and ML, have been producing systems that generate decision trees from data sets: CART [24] and C4.5. These two algorithms are extremely similar. They both build a decision tree by selecting the attributes that compress information the best. An important feature of C4.5 over CART in being used in the context of KDD is that it can run in a *rule-mode* where it generates very comprehensible rules[1], even though they might decrease somewhat precision as compared to the decision tree they come from. It is, or it should be, rather, characteristic of the AI approach[2] to accept losing somewhat on precision for the sake of comprehensibility, while this attitude is not yet admitted in Statistics.
CART and C4.5 perform themselves a discretization of the continuous values, based as well on information compression. Another feature of these programs is the care given to introduce good pruning techniques. In C4.5, the results are commented in terms of this pruning.
Let us underline that, in our experience, we never met a case where C4.5 would perform very badly as compared to a neural network, which is largely confirmed by the ML literature.

2.2.2 Explicit Generalization (KBG, REGAL, SIAO)

These systems are, once more, charaterized by the fact that they deliver explicit discriminating rules out of examples. Comprehensibility is what makes them suitable for KDD. As it often happens, the constrint of comprehensibility eats up computation time, and it is only recently that REGAL and SIAO have been built in order to be able to generate comprehensible rules out of large amounts of exmaples. For instance, KBG is able to deal with a few tens of exmaples, while the new programs leanr from thousands of examples.

---

[1] It can be claimed that a decision tree is also comprehensible. This is true when it is small. In most applications, the obtained decision trees are quite bushy, and rules are much easier to understand, especially those coming from C4.5 in rule-mode because their complexity is reduced as compared to the rules that would be directly drawn from the decision tree.

[2] We have been pushing this idea for a long time (see, for instance, [59]) and it begun to be accepted by most researchers since the beginning of the 90s. It is becoming a kind of banality nowadays.

SIAO learns, using genetic programming techniques [25], rules expressing relations among variables. It uses the star principle developed in AQ for ML [13] and in Diday's "nuées dynamiques" in DA [26, 27]. In SIAO, the seed examples are generalized into rules through a genetic search. This algorithm has been developed in our laboratory. The first system performing this kind of search is REGAL [28] (Giordana et al., 1994) which is much more powerful than the present version of SIAO, but which demands special machines to work, while SIAO works on any UNIX machine (with g++) and on any PC (with Linux). REGAL and SIAO, as ILP systems do, learn relations between entities described by expression with universally quantified variables, as in a clause. More details on SIAO will be given below, in section 4.4.

2.3 The discovery of numerical relations

This particular problem illustrate very well the difference between a Statistical and a ML approach. Regression is a quite well-known technique to associate an equation to a set of points.

In Statistics, regression techniques are used to compute the coefficients of a functional model that fits this model the best to the data. For instance, a least square method tells which is the line, or the polynomial of fixed degree, that is the nearest to some points. The model, here, is the degree of the polynomial, one for a line, for instance. The problem in KDD (and, before, in ML) is the discovery of the model itself since, knowing the model is a prerequisite that most KDD users do not show. This problem might look quite easy, since it is "enough" to try all possible combinations of the base functions of all the variables, to perform a regression for each of the models, and to choose the best one. A combinatory explosion prevents this approach to be interesting. This is why ML has been developing several systems trying to generate the model, in such a way that heuristics drive the choice to the most suitable combination of base functions. The main drawback of this ML approach is that these heuristics, however efficient they might be (which is not always the case), are difficult to understand, and their bias are hard to control. This decreases largely the importance of the old ML approach.

A new technique, based on **Genetic Algorithms** (GAs), performs a very efficient search when the number of base functions is not too high, and does not includes the use of heuristics specially devised for this kind of problem.

3 A NEW FORMULATION FOR SOME CLASSICAL STATISTICAL PROBLEMS

KDD has been modifying the formulation of problems well-known in Statistics, and it developed new approaches, mostly in order to deal with symbolic data, while Statistics has been concentrating on numeric data. Once more, comprehensibility is one important factor in KDD, while precision is the main factor in Statistics. In the course of our text, we shall precise what can be seen as original, or as a reformulation of an old problem.

3.1 Detection of deviations

This work is a reformulation of the statistical deviation measurement, in order to adapt it to symbolic data.

A "deviation" characterizes the fact that a subset of some element shows a behavior which is different from the one of the whole set. Piatetsky-Shapiro and Matheus [29] use this concept to guide the discovery of new relations. The principle of the measurement of deviation follows.

A DB defines standard values of some variables. The DB is divided into regions, called windows or sectors. Within a given window, it might be that the mean value of the considered variable differs largely from normal value. Obviously, many deviations can be detected by different choices of the window under consideration. This forces us to define at once what is an interesting deviation. Three criteria help in the evaluation of interestingness: relation to interesting variables, statistical measurements, and, most of all, the existence of possible strategies in practice, so as to correct the deviation. This last criterion amounts to consider only deviations upon which some action can be taken. For instance, one has no influence on normal pregnancies, while premature birth can be corrected by decreasing the load of the future mother.

The KeFiR system [30] performs this task. It has been applied to the analysis of medical care of the employees of GTE. This system will be made commercially available by GTE.

The computation of deviations

A first hypothesis, without which no computation is possible, if that one knows the normal values of the variables in the DB, together with a causal model, possibly incomplete. Suppose that we want to analyze the deviation of variable (called the *field* in DB terminology) z. If a causal model is known, or discovered by other means (see infra, section 3.4), one knows the causal variables of z. Let, for example, be x and y such variables. Let us then look at sector S.

One gathers the normal values of variable z for each value of x, Norm (x, z), and each value of y, Norm (y, z) in S. One gathers also the relative frequencies of x and y for each of their values, Freq(x) and Freq(y), in S.
The expected value for z is then:

$$z_{exp} = \sum \text{Norm}(x, z) * \text{Freq}(x) + \sum \text{Norm}(y, z) * \text{Freq}(y)$$

In the example below, extracted from [30], (Matheus et al., 1994)

value of x	Norm of z		frequency of x in S
[0, 10]	5		10%
[10, 15]	6		2%
etc.	etc.		
value of y			frequency of y in S
[-3, +3]	2		15%
[3, 5]	1		8%
etc.	etc.		

The value of the experimental frequency is given by:
$z_{exp} = 5*10\% + 6*2\% + \text{etc.} + 2*15\% + 1 * 8\% + \text{etc.}$
Calling z_{mean} the mean of z value in the data, deviation in S is: $| z_{exp} - z_{mean} |$.

3.2 Discovery of causal relations

The problem of detecting causal relations is classical in Statistics. One will find in [31], the description of many existing algorithms. A thorough review of the concept of causality in AI is given in [32].

The importance in KDD of this problem is linked to comprehensibility because most ML systems produce induction based on various measures, all having no link with causality.

Suppose that a ML system invents the rule $A \Rightarrow B$. The domain expert understands this relation following a given semantics, as shown in [33]. For instance, implication might mean causality (as in $\forall$ x [smokes(x) $\Rightarrow$ cancer(x)]), or it can express a property (as in $\forall$ x [crow(x) $\Rightarrow$ black(x)]), and there are many other semantics. Since the inductive system is not based on causality, but on information compression as in C4.5 and CART, it will thus be misunderstood by the domain expert waiting for causality. Depending on data distribution it might happen, for instance, that a theorem such as $\forall$ x [smokes(x) & blond_hair(x) $\Rightarrow$ cancer(x)] is generated for information compression reasons, and the domain expert will try to see why blond_hair can induce cancer, while the system says only that it is easier to distinguish the cases of cancer by looking at the hair color as well.

This explains why we claim that a comprehensibility principle demands that causal dependencies (when the expert is waiting for such dependencies) are first detected, and that the inductive search of possible premises to a theorem be done among causal descriptors, as in [34].

3.2.1 Dependencies among dated facts [35]

This theory is to be applied when the facts one deal with are instanciated, that is to say expressed by the values of discrete attributes, and when they describe time ordered events.

Let x be a conjunction of facts, $x = [x_1 = a_1 \& ...\& x_n = a_n]$, and let y be a fact: $y = b$. As in logistic regression , one says that x and y are dependent if the probability of the occurrences, P, is such that

$$P(y) \neq P(y \mid x).$$

One says that x causes y, if each event $x_i = a_i$ occurs before event y.

These relations are directly understood by a domain expert, such as :

(smokes = YES) $\Rightarrow$ (cancer = YES).

The classical problem of spurious dependencies arises here also. A dependency between z and x is spurious if there exists a y, causal to x, and knowing z does not change the conditional probability of x knowing y, in other words:

$$P(x \mid y) = P(x \mid y, z).$$

3.2.2 Dependencies among non temporally linked variables [34, 36[3]]

---

[3] It was argued during the workshop that Pearl does not support anymore his own theory. Pearl's answer is that he obviously still stands in favor of his work. He acknowledges that this theory guaranties "only" minimality and stability. He adds: "Minimality guarantees that any other causal structure compatible with the data is necessarily more redundant, and hence less trustworthy, than the one(s) inferred. Stability ensures that any alternative structure compatible with the data must be less stable than the one(s) inferred; namely, slight fluctuations in conditions or parameters will render an alternative structure no longer compatible with the data."

This theory is to be applied when one deals with continuous features, not time ordered. The intuition underlying the following definitions is the one of minimal causality models. If there is a causality relationship from x towards y in each of the possible minimal models, then x is said to be causal for y. This choice is obviously open to much criticism. Even if it needs more work, it constitutes an improvement on old approaches.

Let $x, y, z_1, ..., z_n$, be a sequence of n+2 variables. Let us call {S} the set of the subsets of $\{z_1, ..., z_n\}$. One element S_i of {S} is called a context.
We shall say that x and y are conditionally dependent in the context S_i iff
$$P(x, y \mid S_i) \neq P(x \mid S_i) * P(y \mid S_i).$$
A variable x then said to be potentially causal for y if x and y are dependent in all contexts, and if there exists a variable z and a context S_k such that x and z are independent in S_k, while y and z are dependent in S_k.

Two variables x and y are said to be spuriously associated when there exists at least one context into which they are dependent, and that there exists two variables z_i and z_j such that, within the same context, z_i and x, and z_j and y, are dependent while z_i and y, and z_j and x, are non dependent.

A variable x is then genuinely causal for y when x is potentially causal for y and that x and y are not spuriously dependent.

Causal relations among variables constitute a dependency graph which is not yet understandable. For example, with the simpler form of discrete variables, knowing that there is a causality link between variable habit_of_smoking (that can take two values: yes and no), and the variable presence_of_cancer (that can also take two values: yes and no) tells us that four theorems are possible, that is, that smoking or not smoking can cause or not the presence of a cancer. In order to know what is the precise causality, we must use an inductive system that will find the precise rule. In particular, for continuous variables, one must induce the discrete thresholds that play a role in causality. This is done by the classical ML systems, working now only on the causal variables, as we recommended at the beginning of this section.

3.3 Association detection

Associations are very similar to dependence rules as they are discovered by statistical methods (such as in the system SAS, for instance). An association is an uncertain implication, associated to its belief and its support, as explained below. KDD has been concentrating on the study of discrete cases, and when the data are heterogeneous: some are discrete and some are continuous, and when the classical hypotheses on data distribution are not fulfilled.

Actually, the definitions met in KDD literature are specific for the discrete case. The data have to be under the form of a table the columns of which are the values of the attributes, the rows of which are items for which the values of the columns are TRUE or FALSE. For instance, a table describing people by their age and hair color would look like:

	hair_color = blond	hair_color = brown	hair_color = white	age = young	age = old
indiv1	TRUE	FALSE	FALSE	TRUE	FALSE
indiv2	FALSE	FALSE	TRUE	FALSE	TRUE
...	...	...	...	...	...

In most cases, this implies some relatively simple data rewriting. One notices that this approach is presently unable to deal with continuous data.

One understands easily how implications can be induced from such a table: the subsets of rows that are simultaneously TRUE can be seen as being related by an implication. The direction of the implication is determined by the fact that TRUE cannot imply FALSE, while the contrary is possible. More formally, here is a definition of uncertain implication such as given in KDD [37].

Let A be the set of the attribute-values, let A be a subset of A, let a_i be a particular attribute value. Thus, $A = \{a_i\}$, $i = 1, ..., m$, and A is a selection of some of the a_i. Let l_j, $j = 1, ...,$ n, be one of the rows of the relation between the values of the attributes. If for some A, and for a row l_j of the relation, one has $l_j(a_i) = $ TRUE for all a_i of A, one says that $l_j(A) = $ TRUE. Let us still define the set of the rows matching A, m(A) as being the set of the l_k taking the value TRUE for each a_i of A : $m(A) = \{ l_k \mid l_k(A) = $ TRUE$\}$.

Let then A be a subset of A, and A' a subset of its complementary in A. Notice tht all our definitions imply that

$$A \supseteq A, \text{ and } A - A \supseteq A'.$$

With the preceding definitions, we are now able to define a confidence measure, γ, and a support, σ, by stating that the implication A $\Rightarrow$ A' is an association with confidence γ and support σ if :

$$\sum m(A \cup A') \geq n\, \sigma, \text{ and } (\sum m(X \cup Y)) / (\sum m(A) \geq \gamma,$$

where the summation runs on the rows of the table, from 1 to n.

This means that a fraction σ of the rows have value TRUE for all the attributes of A and A', and that at least a fraction γ of the rows having TRUE for all the attributes of A have also a TRUE for all the attributes de A'.

The problem in KDD is to find quickly all possible associations, and among the huge quantities of such relations, to determine those "interesting" to the user. In other words, KDD does not reduces its effort in producing fastest or more accurate algorithms, it looks on ways to take into account as well the goals of the user.

4. KDD GENERATED PROBLEMS

4.1 Inductive queries

As already underlined, ML and BD have been building their own knowledge representations, fitting their own needs. One cannot reconcile them easily. It seems that research is attacking this problem in two ways.

The first one, called semantical optimization of the queries (see for example, [38, 39, 40]) takes into account some domain knowledge to improve the execution time of the query. The main difference between the semantic approach and other time improving techniques is that semantical optimization uses rules, and that these rules are automatically generated for the query when it needs it. The query is thus inductively modified on the fly.

The second one addresses directly the problem of inductive queries. Two large systems, namely DBMiner [41] and DataMine [42], introduce specific extensions to SQL in order to perform KDD. This topic seems very important since it links directly DB and KDD. It is interesting to notice, however, that only a short paper [43] describes DMQL, the query

language of DBMiner, which, as stated by the authors, is not "complete by any standard ... it may serve as an interesting example for further discussion."
DMQL adds three functionalities to SQL.

The first one is the possibility to call a DM system within the query. One can well imagine an extension of this by allowing a call to any DM system, not only to those included in DBMiner. This will be possible when the field matures more, and that mining standards are defined.

The second one is the use of background knowledge under the form of concept hierarchies. These hierarchies will be used in order to search, or to deliver answers at different levels of generality. It is clear that background knowledge is more than concept hierarchies, and all kinds of implications should be authorized as well, for instance causality relationships.

The third one is the specification of thresholds under which the mined information is deemed not interesting. DMQL offers three such thresholds: significance (or support: how large is the support of pattern), confidence in a rule, redundancy (how much a rule must differ from the existing ones to be accepted as a new rule). This topic as well should be discussed in the KDD community and a normalized set of thresholds will be accepted when the different possible approximations will be correctly defined.

4.2 Attribute focusing

A new method, based on similar needs has been proposed by Bhandari [44]. It is called Attribute Focusing. Presently, it makes use only of statistics on the occurrences of the values of the attributes.

Let q_1 and q_2 be two attributes, and let us consider the case where $q_1 = u$ and $q_2 = v$. Then, several expressions can measure the degree of association of these two values. Bandhari introduced the measure :
$$I_{bandhari}(q_1 = u, q_2 = v) = \| P(q_1 = u, q_2 = v) - P(q_1 = u) * P(q_2 = v) \|.$$

One could obviously use as well:
$$I(q_1 = u, q_2 = v) = P(q_1 = u, q_2 = v) / (P(q_1 = u) + P(q_2 = v)).$$

Let q_1 be a variable of obvious interest to the user, then focusing on q_2 and q_3 is performed by sorting q_2 and q_3 such q_2 is strongly associated to q_1, and q_3 weakly associated to q_1. The user's attention is focused on these values.

Much more research is still necessary on this topic, which has been recognized of such importance in KDD, that the first KDD theoretical paper we know of [45], is devoted to the study of the influence of such focusing measures on the complexity of association mining.

5. EXAMPLES OF KDD RESEARCH IN OUR GROUP

Let us now describe four KDD problems that are dealt with in our research group.

5.1 Some modifications to C4.5

This research is done at the company EPCAD by A. Marret.

C4.5 minimizes the choice of the best attribute locally only, therefore the whole tree might not be optimal, and direct intervention of the user can push C4.5 in the direction of optimality. This is why we added a few parameters to it, especially in view of allowing the user to impose some constraints on the growth of the tree.

We thus added different functions for turning off, at some of its points, C4.5's process. They allow a user to impose some partial control on the process of tree induction. The different knowledge types that we have introduced with the new options is handled by several modules added to C4.5. These options perform the following tasks :

a - The user can decide what will be the head of the decision tree; normal C4.5 follows after this head. An editor helps to assign the nodes one wants to keep at each level of the tree.

b - Normally, C4.5 decides itself of all thresholds while performing its discretization of continuous variables. In our version, the user can decide of some of these thresholds.

c - C4.5 considers that all variables are independent. We introduced the possibility to take into account links between attributes given by a relation graph between them. An graph manager enables to write down these relations, to watch them, and to put them in files.

d - recomputing missing values by a procedure based on the frequency of the known values.

e - possibility to create new attributes out of the existing ones. A relation manager helps the user to enter this information.

f - possibility to interact directly in the building of the decision tree:

to forbid temporarily the use of an attribute
to stop some branches
to prohibit the formation of a cycle for some continuous attributes.

g- if by any means, the user knows the existence of a numerical relationship in the data, new variables expressing these relations can be added. This mimics partly the use of an oblique decision tree, and allows us to link C4.5 to the "relation discoverer" described below in section 4.3.

5.2 Combining existing systems

This research is done at the company EPCAD in order to create a simple inexpensive KDD software.

Let us see how supervised and non supervised learning can be combined. These two paradigms are linked, as section 3 shows. They should be combined in order to yield better results, especially when the classes are either unknown or badly defined. DA provides systems that perform this combination, as for instance the system SICLA [46]. This system applies first a classification algorithm, then it determines the most significant features, and finally provides an intentional description of the classes. In ML, pioneering systems such as EPAM [47] and UNIMEM [48] performed this feast as well. Some of the systems learning from data expressing relations, such as MOBAL [49], and KBG [22, 23] perform as well supervised and non supervised learning. KBG uses even a preliminary classification in order to avoid the combinatory explosion due to the many possible matchings.

Many researchers have been using such a combination of supervised and non supervised learning on particular data. Nevertheless, they never have analyzed this problem so as to propose general solutions. As a first integration step, we propose to use the following scheme:

We combine these systems in accordance to the following scheme.

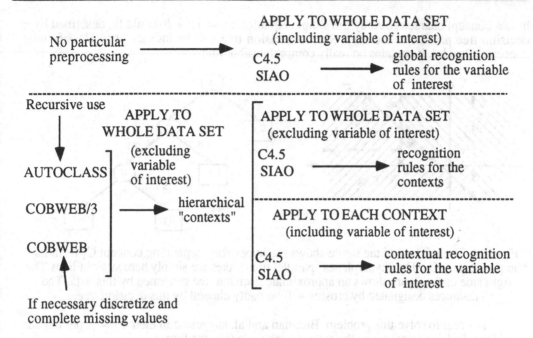

Figure 3. The three possible modes of functioning. The "variable of interest" is defined, as in Data Analysis, as the feature upon which the rules must conclude.

- Mode 1

The rule generators can be obviously applied to the whole data set, including the variable of interest. This is the classical ML behavior which yields recognition rules. When the data are very diverse, this approach yields complex and not very precise rules.

- Mode 2

This mode uses first the cluster generators without the variable of interest. Thus creates classes that are independent of the problem at hand since the variable of interest is not used. This is why we call: "contexts", the classes thus generated. One then uses the rule generators to generate rules that recognize the contexts. Obtaining a rule is very important since its premise is a new feature which describes the relative context.

- Mode 3

In this mode, one applies the rule generators within each context, using the features one starts from, but also the new features generated in mode 2. In the case of extremely large data sets, this operation constitutes a very natural way of focusing.

5.3 Computing hyperplanes separating the data

This research is done at LRI by A. Brunie-Taton and A. Cornuéjols.
The systems C4.5 and CART generate decision trees that are said to be *parallel to the axes* because the test at each node of the tree is the equation of a hyperplane parallel to the axes in the space of the features values. This fact can prevent discovering concepts that could otherwise be described by a simple combination of features. Figure 4 below shows

how a concept characterized by the simple equation : $y = a x + b$, could be described by a decision tree parallel to the axes. Such a decision tree will be inefficient to classify new concepts. Besides, it contains no really comprehensible rule.

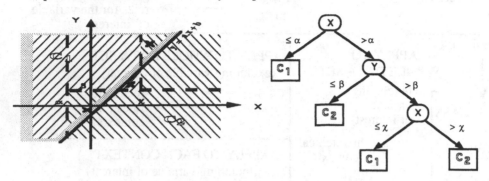

Figure 4. The left side of the figure shows the hyperplane separating concept C_1 from its complementary C_2. The "hyperplanes" parallel to the axes are simply here straight lines.The right side of the figure shows an approximate decision tree generated by this data. The instances designated by crosses will be badly classed by this decision tree.

In order to solve this problem, Breiman and al. suggested in their 1984 paper the use of oblique decision trees, where the tests at each node take the form :

$$\sum_{i=1}^{n} a_i x_i + a_{n+1} > 0$$

where $a_1,...,a_{n+1}$ are real valued coefficients. It is obvious that oblique decision trees are more efficient and precise when the concepts to recognize are defined by a polygonal partition of space that will be easier to approximate by non linear hypersurfaces. Computational complexity is however an important issue. Finding an optimal decision tree parallel to the axes is already NP-hard, which is still increased by the search, at each node of the tree, of an optimal linear combination of the axes. This why, after the pioneering work of Breiman, one had to wait for the 90s to obtain practical methods for the induction of oblique decision trees [50, 51, 52, 53, 54]. All these try to find a good tradeoff between computational efficiency and optimal choice of tests at each node. Their methods are easily stuck in local minimum, this is why we prefer using GAs.

On the other hand, as we already insisted upon, optimal precision is not the ultimate KDD goal, which includes comprehensibility as well. The least one can say is that oblique decision trees generate very obscure results: the coefficients of linear combination of axes that has been chosen at each node. This is why we propose to stop working on oblique decision trees, and to find rather the corresponding linear combination of data that will produce the same effect.

Suppose, for instance, that an oblique node is needed because, say, variables y and x are linearly dependent, as expressed by the equation
    $y = a x + b$.
It is quite obvious that $(y - b)/x = a$ is a constant, and, as a consequence, $(y - b)/x$ is a variable that separates x and y no longer by the line $y = a x + b$, but by the line of equation "a", parallel to the axes. Introducing such a variable is equivalent to using an oblique decision tree, but it preserves comprehensibility, as long as this new variable, $(y - b)/x$, is intelligible to the user.

The principle of this research is parallel to some research in Statistics, but cannot be confused with it, since regression will find hyperplanes the nearest to a given number of points, while we look for an hyperplane that separates data classes for the variable of interest. Moreover, these hyperplanes must be as much "central" as possible relative to the classes they separate. Our experimental results [55] (Brunie-Taton & Cornuéjols, 1996) show clearly that DA classical distance measures lead to very bad results as compared to those obtained using information compression.

Notice also that, in case of really too imprecise results, one can then accept to somewhat lose on comprehensibility by introducing models more complex than hyperplanes, for instance an hypersphere. Genetic programming allows to test relatively quickly such complex models.

5.4 Quickly generating relational formulas

This research is done at LRI by S. Augier and at University of Tours by G. Venturini. It lead to the system SIAO that we described shortly above. As we said, REGAL from Giordana is the main other system able to perform this kind of work.

An important difference between SIAO and REGAL [28, 56] [(Giordana, Saitta & Zini, 1994 Giordana & Saitta, 1993;) is that REGAL demands a model of the rules to be discovered. For instance :

Color(X,[red, blue, *]) & Form(X,[square, triangle, *]) & Not-exists Y [Color(Y, [red, blue, *]) & Distance(X, Y,[0, 1, 2, 3, *])]

In this example, Color (X, [red, blue; *]) means that there exists an entity, named X, such that its colors can be red, or blue, or neither red nor blue. This gives only 6 possible values to the list [red, blue, *] which can only take the values [1,1,1], [1,1, 0], etc. Using such predefined patterns reduces surely the combinatorics. It nevertheless limits severely also the inventive abilities of the algorithm.

Inversely, SIAO asks for no preliminary model, and it uses the form of the data to generalize directly, exactly as KBG and AQ do. Moreover, SIAO accepts missing, undefined, or non relevant values. Its weaknesses as compared to REGAL are computation time, and that it cannot discover existentially quantified relations, nor negations.
In SIAO, one of the examples is used as a seed, its expression defines the model one starts from, and it is generalized by using the classical generalization techniques such as replacement of constants by variables, climbing the generalization tree, etc., exactly as in AQ.
An evaluation function decides which is the best rule generated by the seed example. This rule being found, the examples it covers are discarded, and process starts again with the examples left, until no example is left, as in AQ.

5. CONCLUSION

Definition :
    KDD is a new field of research. As compared to its more aged parents, it is characterized by the fact that it retrieves comprehensible and interesting knowledge from data by using (also) inductive techniques, where *the data is used as provided by the user.* Its goal is measuring the interestingness and comprehensibility of the generated knowledge.

As announced in the introduction, the present instance of this definition is still incomplete since there are very few existing KDD systems able to measure interestingness, and comprehensibility is hard to measure properly. The body of this paper shows however that hopes to improve in these directions are not at all chimerical.

This definition uses three badly defined terms, namely **knowledge**, **comprehensible**, and **interesting**. Let us attempt to show how they are defined in a quite rigorous fashion in KDD.

Before defining what knowledge means in KDD (obviously, we do not intend to define what knowledge is in general), let us define the intermediate concept of a **piece of knowledge**.

A **piece of knowledge** is a sentence in a predefined language which shows the following properties:

1 - The sentence is twofold. One part is the "knowledge" itself (for instance, an implication and a coefficient of belief in this implication), and the other part is "meta-knowledge". This meta-knowledge describes the origin of the knowledge. For instance, an implication may have been inferred from statistical relations in data, or alternatively from a specific experimentation in the real world. Seemingly in opposition to us, people in Knowledge Acquisition (KA) ask traditionally that this meta-knowledge describes how to use knowledge (for instance by setting the contexts into which an implication is *interesting to use*, and those into which it is *useless*, even though the implication is *valid* everywhere). It is obvious that KDD has no access to the models of knowledge to use, and it cannot fulfill this KA request. Nevertheless one should be able to infer it from knowledge. We thus suppose here that from the relevant pieces of knowledge, their origin, and the goals of the agent (as seen in property 2, below), one should be able to deduce the way to use the knowledge. This is a strong supposition which might be recognized later as erroneous. In that case, our definition would have to be changed so as including more than the origin of knowledge, such that this meta-knowledge, together with user's goals, are enough to know how to use knowledge

2 - There exists at least one actor, a living being or a machine, that will behave differently in the world depending whether it possesses this piece of knowledge or not. The degree of relevance to a particular action of a piece of knowledge is often the topic of many discussions. This agent has goals that determine the use of knowledge, together with its origin.

As an example of the consequences of this definition, let us briefly discuss whether or not animals can show some "knowledge". In our definition, it is clear that we demand a language plus an action in order to recognize knowledge. A dog which barks in a special way (sentence in a language) and sniffs at a place (origin of the knowledge), then digs out a hidden bone (discovery), will be said to have shown, in a KDD sense, a piece of knowledge of the presence of a bone. This is possible because we do not ask a piece of knowledge to be grounded.

KDD **knowledge** can then be defined as a set of pieces of knowledge showing little enough incoherence and enough redundancy so as to execute successfully precise actions in the world, while using imprecise pieces of knowledge.

In practice, most existing KDD tools do not provide knowledge but pieces of knowledge. The research program of KDD, nevertheless, includes the problem of putting these pieces together in order to present knowledge to their user.

Again in practice, while all DM tools include the truth-value of the pieces of knowledge they find, they rarely give a judgment on the origin of the knowledge. A cornerstone for KDD is to deal with this problem, touched upon only by those concerned with interestingness, and who do not reduce it to the measure of truth-value. We do not know of a single system that evaluates the worth of the data before mining it. Our definition of KDD knowledge includes obviously the need for such functionality, which has been up to now fulfilled by a discussion between DM and field specialists.

Interesting can be better understood in view of the above definitions: an interesting piece of knowledge is one for which
1 - truth value is high enough
2 - the origin of the information is well described
3 - relevance value is high enough

One must confess that interestingness is still almost completely user-defined in the present state of the art of KDD, except in the pioneering works of Piatetsky-Shapiro and Bandhari presented in section 3.2 and 3.3.

Comprehensible means expressed in the user's language and with the user's semantics. Since most people understand what they see, visualization is an important part on comprehensibility. Also in many cases, people understand better what is short, thus the length of each piece of knowledge is very relevant to its comprehensibility. For instance, a user might say that rules with premises containing more than three predicates are not comprehensible. The nature of the predicate used is also very relevant to comprehensibility. For instance, even if he understands perfectly the rule, a warrior may claim that rules discriminating among tanks by the form of their chains are not comprehensible, while the mechanics in charge of repairing the tanks could find it perfectly understandable. Interestingness is obviously goal oriented, but it seems that comprehensibility is also: uninteresting is harder to understand.

We do not include in our definition, as [3] does, the fact that the discovered knowledge should be implicit in the data, and previously unknown. Adding this would limit the goals of the user who might, for instance, happy (that would be a special kind of comprehensibility) to find, together with unknown knowledge, something he knew already.

The above definition implies a large number of research themes. One particularly striking example is the search for causal links, and its subsequent use for discovering rules that make more sense to the user since the premises of the rules are indeed causal, as users often expect.

In this paper, we insisted on the fact that KDD, as all other fields, possesses its own logics, by which it can judge what is belonging to its field and what is not. Up to now, the emphasis has been put on its importance in Industry [57] and for other scientific fields [58] (Fayyad et al., 1996b). The conjunction of the two properties, namely that it is self-defined and that it is useful, makes it socially well-grounded, which is not so often the case for many other fields that develop under their own logic instead of a good balance between their own logic and external requests.

When the data is numerical, symbolic (= discrete + semantics different from numbers), heterogeneous symbolic-numeric, textual, multi-media, or commented DB, in all these cases, we should produce efficient tools, clearly documented. This is not yet true, but it is one of the main themes of KDD research.

It is also another research field to systematically exploit the **users' knowledge**, and to determine which of this knowledge is useful for given DM tools.

At a longer range, KDD systems will also have to provide **explanations** to the users on why such or such discovery has been performed. This demands a perfect mastership of induction bias, one of the most advanced research topics of ML.

REFERENCES

1. Frawley W., Piatetsky-Shapiro G, Matheus C.: Knowledge Discovery in Databases: An Overview, AI Magazine, Fall 1992. Reprint of the introductory chapter of Knowledge Discovery in Databases collection, AAAI/MIT Press, (1991).

2. Kodratoff Y.: Foreword of the guest editor: The Comprehensibility Manifesto, AI Communications, 7, (1994).

3. Fayyad U., Piatetsky-Shapiro G., Smyth P.: Knowledge Discovery and Data Mining: Towards a Unifying Framework, Proc. 2nd International Conference on KDD & DM, Simoudis E. and Han J. (Eds.), AAAI Press, Menlo Park CA, (1996), 82-87.

4. Cheeseman P., Stutz J.: Bayesian Classification (AutoClass): Theory and Results, in: Advances in Knowledge Discovery and Data Mining, U. M. Fayyad, G. Piatetsky-Shapiro, P. Smyth, R. Uthurusamy (Eds.), The AAAI Press, Menlo Park, 1996.

5. Turing A. M.: Computing Machinery and Intelligence", Mind 59 (1950), 433-460.

6. Searle J. R.: Minds, brains & science, Penguin books, London 1984.

7. Searle J. R., Scientific American 262 (1990), 26-31.

8. Draganescu M.: L'Universalite' ontologique de l'information (Ontological Universality of Information), preface & notes by Y. Kodratoff, Bucharest, Editura Academiei, 1996 (also available at http://www.racai.ro/books/draganescu). (In French). See also Draganescu M.: Information, Heuristics, Creation, in : Artificial Intelligence and Information, Control Systems of Robots, Plander I. (Ed.), Elsevier, Amsterdam 1984, 25-29.

9. Brachman, R.J., Anand, T.: The Process of Knowledge Discovery in Databases: A First Sketch, in: Proc. KDD'94, Seattle, (1994), 1-11.

10. Wirth R., Reinartz T. P.: Detecting Early Indicator Cars in Automotive Database: A Multi-Strategy Approach, in: Proc. 2nd International Conference on KDD & DM, Simoudis E. and Han J. (Eds.), AAAI Press, Menlo Park CA, 1996, 76-81.

11. Lindner G., Morik K.: Coupling a relational learning algorithm with a DB system, ECML workshop on ML, Statistics, and KDD, Heraklion, 1995.

11. Riddle P., Segal R., Etzioni O.: Representation design and brute-force induction in a Boeing manufacturing domain, Applied Artificial Intelligence 8 (1994), 125-148.

12. Kodratoff Y., Vrain C.: Acquiring first-order knowledge about air-traffic control, Knowledge Acquisition, 5 (1993), 1-36.

13. Michalski, R.S.: A Theory and Methodology of Inductive Learning, in: Machine Learning: An Artificial Intelligence Approach, Vol. 1, R.S. Michalski, J.G. Carbonell, T.M. Mitchell (Eds.), Morgan Kaufman, Menlo Park CA 1983, 83-134.

14. Benzecri J. P. (with many co-authors) L'analyse des données, Dunod, Paris, 1973. (In French)

15. Fisher D.: Knowledge acquisition via incremental conceptual clustering, Machine Learning 2 (1987), 139-172.

16. Gennari J. H., Langley P., Fisher D.: Models of incremental concept formation, Artificial Intelligence 40 (1989), 11-61.

17. McKusick, K., Thompson, K.: Cobweb/3: A portable implementation (Technical Report FIA-90-6-18-2). Moffett Field, CA.: NASA Ames Research Center, Artificial Intelligence Research Branch, 1990.

18. Ketterlin A., Korczak J. J.: Concept formation in complex domains, in: Proc. ECML, Vol. 784 Springer-Verlag's LNCS, Berlin 1994.

19. Ketterlin A., Gançarski P., Korczak J. J.: Conceptual clustering in structured databases: a practical approach, in: Fayyad U. M., Uthurusamy (Eds.) Proc. KDD'95, Montreal, AAAI/MIT Press 1995.

20. Quinlan J. R.: C4.5: Programs in Machine Learning. Morgan Kaufmann, Menlo Park CA 1992.
21. Augier S., Venturini G., Kodratoff Y.: Learning first order logic rules with a genetic algorithm, in: Fayyad U. M., (Eds.) Proc. KDD'95, Montreal, AAAI/MIT Press 1995.
22. Bisson G.: Learning in FOL with a similarity measure, in: Proceedings of AAAI, San Jose, California, 13-17 July 1992.
23. Kodratoff Y, Bisson G.: The epistemology of conceptual clustering: KBG, an implementation", Journal of Intelligent Information System, 1(1992) 57-84.
24. Breiman L., Friedman J., Olshen R., Stone C.: Classification and Regression Trees, Wadsworth International Group 1984.
25. Koza J.: Genetic Programming: On the Programming of Computers by Means of Natural Selection, The MIT Press 1992.
26. Diday E.: The dynamic clusters method in non hierarchical clustering, International Journal of Computer Sciences, 2 1973.
27. Michalski R. S., Diday E., Stepp R. E: A recent advance in Data Analysis: Clustering objects into classes characterized by conjunctive concepts, in Progress in Pattern Recognition, Kanal and Rosenfeld (Eds), 1982.
28.Giordana A., Saitta L. and Zini F.: Learning disjunctive concepts by means of genetic algorithms, in: Proc. 11th International Conference on Machine Learning, 1994, 96-104.
29. Piatetsky-Shapiro, G. Matheus, C.J.: The interestingness of deviations, in. Proc. KDD 94, 1994, 25-36.
30. Matheus, C.J., Piatetsky-Shapiro, G., McNeill, D.: An Application of KeFiR to the Analysis of Healthcare Information, in Proc. KDD 94, pp. 441-452, 1994.
31. Spirtes P., Glymour G., Scheines R.: Causation, prediction and search, in: Lectures Notes in Statistics-81, Springer-Verlag, Berlin 1993.
32. Findler N. V., Bickmore T.: On the concept of causality and a causal modeling system for scientific and engineering domain, Applied Artificial Intelligence 10 (1996), 455-487.
33. Kodratoff Y.: Induction and the Organization of Knowledge, in: Machine Learning: A Multistrategy Approach, Tecuci G. & Michalski R. S. (Eds.), pages 85-106. Morgan-Kaufmann, San Francisco CA, 1994.
34. Esposito F., Malerba D., Ripa V., Semeraro G.: Discovering Causal Rules in Relational Databases, in: Cybernetics and Systems'96, R. Trappl (Ed.), Austrian Soc. for Cyber. Studies, Vienna, Austria 1996, 943-948.
35. Pavillon G.: ARC II: a System for Inducing and Simplifying Dependence and Causal Relationships, in: Cybernetics and Systems'96, R. Trappl (Ed.), Austrian Soc. for Cyber. Studies, Vienna, Austria 1996, 985-990.
36. Pearl J., Verma T. S., A theory of inferred causation, in: Allen J. A., Fikes R., Sandewall E. (Eds.), Principles of Knowledge Representation and Reasoning, Morgan Kaufmann, San Mateo CA 1991, 441-452.
37. Mannila H., Toivonen H., Verkamo A. I.: Efficient algorithms for discovering association rules, in: Proc. KDD'94, pp. 181-192, Seattle, July 1994.
38. Siegel M. D., Sciore M., Salveter S.: A method for automatic rule derivation to support semantic query optimisation, ACM Trans. on DBS, 17 (1992), 530-600.
39. Hsu C., Knoblock C. A.: Rule induction for semantic query optimisation, Proc. 11th IMLC, 1994.
40. Sayli A., Lowden B.: The use of statistics in semantic query optimisation, in: Cybernetics and Systems'96, R. Trappl (Ed.), Austrian Soc. for Cyber. Studies, Vienna, Austria 1996, 991-996.
41. Han J., Fu Y., Wang W., Chiang J., Gong W., Koperski K., Li D., Lu Y., Rajan A., Stefanovic N., Xia B., Zaiane O. R.: DBMiner: A System for Mining Knowledge in Large Relational Databases, in: Proc. 2nd International Conference on KDD & DM, Simoudis E. and Han J. (Eds.), AAAI Press, Menlo Park CA 1996.

42. Imielinski T., Virmani A., Abdulghani A.: DataMine: Application Programming Interface and Query Language for Data Mining, in: Proc. 2nd International Conference on KDD & DM, Simoudis E. and Han J. (Eds.), AAAI Press, Menlo Park CA 1996, 256-261.
43. Han J., Fu Y., Wang W., Koperski K., Zaiane O.: DMQL: A Data Mining Query language for Relational Databases, in: SIGMOD'96 Workshop. on Research Issues on Data Mining and Knowledge Discovery (DMKD'96), Montreal, Canada, June 1996.
44. Bhandari, I.: Attribute focusing: Machine-Assisted Knowledge discovery Applied to Software Production Process Control, Knowledge Acquisition 6 (1994), 271-294.
45. Mannila H., Toivonen H.: On an algorithm for finding all interesting sentences, in: Cybernetics and Systems'96, R. Trappl (Ed.), Austrian Soc. for Cyber. Studies, Vienna, Austria 1996, 973-978.
46. Ralambondrainy H.: An interactive system of classification : SICLA, in: H. Caussinus et al. (Eds.), Proceedings in Computational Statistics, COMPSTAT 82, Physica-Verlag, Wien 1982, 225-225.
47. Feigenbaum E. A.: The simulation of verbal learning behavior, in: Computers and Thought, Feigenbaum E. A & Feldman J. (Eds.), McGraw-Hill, N.Y. 1963.
48. Lebowitz M.: Experiments with incremental concept formation/ UNIMEM, Machine Learning 2 (1987), 103-138.
49. Morik, K., Wrobel, S., Kietz, J. U. and Emde, W.: Knowledge Acquisition and Machine Learning - Theory, Methods, and Applications. Academic Press, London 1993.
50. Utgoff P., Brodley C.: An incremental method for finding multivariate splits for decision trees, in: Proc. of the Seventh International Conference on Machine Learning (ICML-90), Morgan Kaufmann, Los Altos, CA 1990, 58-65.
51. Heath D, Kasif S., Salzberg S.: Learning oblique decision trees, in: Proc. of the 13th International Joint Conference on Artificial Intelligence, Morgan Kaufmann 1993, 1002-1007.
52. Murthy S., Kasif S., Salzberg S. & Beigel R.: OC1: Randomized induction of oblique decision trees, in: Proc. of the Eleventh National Conference on Artificial Intelligence, MIT Press, Washington D.C. 1993, 322-327.
53. Murthy S., Kasif S. & Salzberg S.: A System for Induction of Oblique Decision Trees, Journal of Artificial Intelligence Research 2 (1994), 1-32.
54. Brodley C. & Utgoff P.: Multivariate decision trees, Machine Learning 19 (1995), 45-77.
55. Brunie-Taton A. & Cornuéjols A.: Classification en Programmation génétique, in: Proc. of the 11th Journées Françaises d'Apprentissage (JFA-96), Sète, France, May 8-10, 1996. (In French)
56. Giordana A., Saitta L., Regal: an integrated system for learning relations using genetic algorithms, in: Proceedings of the Second International Workshop on Multistrategy Learning, R.S. Michalski and G. Tecuci (Eds.) 1993, 234-249.
57. Piatetsky-Shapiro G., Brachman R., Khabaza T., Kloesgen W., Simoudis E.: An Overview of Issues in Developing Industrial Data Mining and Knowledge Discovery Applications, in: Proc. 2nd International Conference on KDD & DM, Simoudis E. and Han J. (Eds.), AAAI Press, Menlo Park CA 1996, 89-95.
58. Fayyad U., Haussler D., Storloz P. "KDD for Science Data Analysis: Issues and Examples," Proc. 2nd International Conference on KDD & DM, Simoudis E. and Han J. (Eds.), AAAI Press, Menlo Park CA 1996, 50-56.
59. Kodratoff Y., "Is AI a sub-field of Computer Science or AI is the Science of Explanations", in: Progress in Machine Learning, I. Bratko & N. Lavrac (Eds.), Sigma Press, Wilmslow 1987, 91-105.

MACHINE LEARNING

MACHINE LEARNING:
BETWEEN ACCURACY AND INTERPRETABILITY

I. Bratko
Ljubljana University, Ljubljana, Slovenia

ABSTRACT

Predictive accuracy is the usual measure of success of Machine Learning (ML) applications. However, experience from many ML applications in difficult domains indicates the importance of *interpretability* of induced descriptions. Often in such domains, predictive accuracy is hardly of interest to the user. Instead, the users' interest now lies in the interpretion of the induced descriptions and not in their use for prediction. In such cases, ML is essentially used as a tool for exploring the domain, to generate new, potentially useful ideas about the domain, and thus improve the user's understanding of the domain. The important questions are how to make domain-specific background knowledge usable by the learning system, and how to interpret the results in the light of this background expertise. These questions are discussed and illustrated by relevant example applications of ML, including: medical diagnosis, ecological modelling, and interpreting discrete event simulations. The observations in these applications show that predictive accuracy, the usual measure of success in ML, should be accompanied by a criterion of interpretability of induced descriptions. The formalisation of interpretability is however a completely new challenge for ML.

1 Introduction

In Machine Learning (ML), the usual criterion of success has been considered to the *predictive accuracy*.

However, observations from many ML applications show that the user's main interest was in the *interpretation* of the descriptions induced from examples. These observations come from applications in various domains. In this paper we look at some of such applications that come from ecological modelling [6], predicting the mutagenicity of chemical compounds [12], assessing river water quality [3], modelling the deer population dynamics [13], and analysis of discrete event simulation results [9]. For example, Boris Kompare, a user of ML techniques, studied in his thesis the growth of algae in the Lagoon of Venice and the biodegradability of chemicals [6]. The main contribution of the thesis consisted in discovering, with machine learning, of interesting patterns in the environmental data. Although the predicitve accuracy of the induced descriptions was very limited, they were of interest due to their expert interpretations. The interpretations then made the learning results useful as a source of ideas about the domain of investigation. Without the in-depth interpretations of learning results, the thesis would be of little interest.

Comments from system modelling and simulation experts are also quite revealing in respect of experts' interests. Simulation results are rarely used really for making for predictions. Instead, models are studied and simulations done to deepen the expert's *understanding* of the problem.

All this experience indicates that in machine learning, the predicitve accuracy criterion has been perhaps over emphasised, and the interpretability criterion has been under emphasised. Probably the reasons for this situation are two fold. First, there are well understood and accepted methods for estimating accuracy. They include such methos as cross-validation and statistical significance tests. On the other hand, there is no known formalised method of estimating the *degree of interpretability*. There is no measure of interpretability. It is not even quite clear what aspects a *qualitative* assessment of interpretability should involve.

In the remainder of this paper we look in more detail into some examples of ML applications and discuss the aspects of interpretability. The examination of these cases reveals that some "built-in" assumptions in machine learning should be reviewed.

Much more attention should be paid to criteria quite different from the usual predictive accuracy, or similar accuracy-based criteria such as information score [7]. Observations in these example applications also indicate some "ingredients of interpretability". More generally, they indicate what is needed for induced descriptions to qualify as knowledge. This discussion points towards a number of open problems for new research in machine learning.

2   Observations in medical diagnosis

Here we consider two studies in which, among other aspects, the meaningfulness of induced descriptions to medical doctors was also studied.

2.1   Rheumatology

In an application of ML to diagnostic data in rheumatology [11], we were interested in medical specialist's comments regarding the induced decision trees. A brief summary of this experiment is as follows: The expert commented that many of the induced trees were "unnatural", and for this reason not really satisfactory. This was despite the fact that the measured diagnostic accuracy of the trees was significantly above the measured accuracy of medical experts themselves in this domain. One specific expert's comment was that "the trees did not tell enough, they should give more information". The expert was told that the trees had in fact been pruned, because of noise in the learning data, in order to optimise their accuracy. So any further "information" in the trees would be redundant with respect to diagnostic accuracy. Nevertheless, the expert's opinion remained unchanged. He was then encouraged to modify the trees so that they would become more acceptable from his point of view, and would also include the "missing information". After the expert's modifications, the trees became larger and their diagnostic accuracy decreased slightly. In spite of this decrease in accuracy of the trees, the expert now liked them and found them clearly preferrable to the ones originally induced.

This experiment indicated that the expert was not particularly interested in the predictive accuracy and minimal description length – two dominating criteria in ML.

2.2   Cardiac arrhythmias

The second relevant experience in a medical application comes from the domain of cardiac arrhythmias. In the KARDIO project [1], a complete knowledge base about the relations between cardiac arrhythmias and the features of the corresponding electro cardiograms (ECGs) was compressed, using machine learning programs, to a much more compact and still correct set of rules. In the generation of these rules, the condition was that the rules were correct and complete and of minimal length. Among the induced rules there were also definitions, automatically generated, of well known physiological mechanisms. These mechanisms are also described in the medical literature. It was then very illustrative to compare machine-synthesised rules with those proposed by the medical experts in medical books.

Here we consider an example of such definitions, one that describes the heart disorder called AV block of the third degree. This defect arises when the atria and the ventricles are electrically disconnected, so the electrical signals that generate the heart beat cannot get from the atria to the ventricles. The machine-generated rule (straightforwardly transcribed into the usual medical language) is:

"If the rhythm of the ECG is regular and the QRS complexes are independent of the P waves then there is complete AV block". This conditional statement holds generally, that is for *any combination* of heart disorders that includes the complete AV block.

The corresponding descriptions in the classical medical books about the heart physiology by Goldman and Phibbs state:

Goldman: "In this condition the atria and ventricles beat entirely independently of one another. ... . The ventricular rhythm is quite regular but at much slower rate (20 to 60)."

Phibbs: "1. The atrial and ventricular rates are different: the atrial rate is faster; the ventricular rate is slow and regular. 2. There is no consistent relation between P waves and QRS complexes."

All the descriptions are very similar, except for the linguistic diferences. Both humans' descriptions say the same. The machine-generated one is a logical subset of the human's descriptions. The only difference is that both human experts mention the slow rate of the ventricles, whereas the machine-generated rule does not mention this feature. Obviously, since the machine-generated description is minimal, this feature is, strictly speaking, redundant. It could be speculated that the humans, who could not check this redundancy against the many possible combinations of heart disorders, added this feature just to be safe. However, it is much more likely (as also confirmed by a medical doctor) that they added this redundancy deliberately as an additional discriminating feature even if not strictly needed. But this extra feature would still be useful as a *confirmation* of the diagnosis.

The comparison of other rules and medical descriptions showed that the human definitions were often quite incorrect if really taken literally. In such cases the correct, machine-generated rules were much more complex. Again, the medical experts' simplifications seemed to be quite deliberate. The simplifications relied on a justified assumption that a knowledgeable reader would be able to adjust the simplified descriptions to the details of exceptional special cases.

This experiment indicates that human experts sometimes find redundancy useful and preferrable to minimal-length descriptions. The opposite is also often true: the human experts are quite happy to sacrifice (strict) correctness in order to gain simplicity. It seems that simplifications are justified by unstated assumptions which the reader should be able to make himself on the basis of his background knowledge.

3   Ecological modelling

Here we discuss two cases of relevant experience: first, ML application to the modelling of the growth of algae in the Lagoon of Venice, and second, the modelling of forest development depending on the deer population in the forest.

3.1   Growth of algae

In the Venice Lagoon study, measured data series were used as examples for ML programs to model the growth of algae depending on a number of parameters. The measured parameters in these experiments were: the biomass of algae per square meter, temperature of water, and the concentrations of nitrogen in two forms, phosphorus and disolved oxygen. Two learning programs were used: Retis [5] and GoldHorn [8]. Retis induces regression trees, and GoldHorn induces differential or difference equations from data series.

The models induced by both Retis and GoldHorn were of low accuracy. This was not unexpected since the measured data was scarce and very noisy. The measurement error was estimated by the expert to be about 50%. Still, both regression tree models and difference equations models induced were of interst to the domain experts. Their own very complex differential equations models, resulting from some ten years of modelling efforts in the traditional style, also have very limited predictive accuracy. One of the models induced by GoldHorn had very similar accuracy, although the GoldHorn model was much simpler than the experts' model.

It was very informative to follow the expert interpretation of regression tree models induced by Retis. Although regression trees predict numerical values, the interpretation was almost entirely *qualitative*. The experts were interested in what happens if a certain parameter decreases or increases: How will such changes effect the biomass of algae? For example, how does the concentration of phosphorus in water affect the biomass of algae in time?

A relation was found in the Retis models that appeared rather counter intuitive to at least one of the experts. A surprizing pattern was repeatedly occuring in the induced regression trees. It is qualitatively shown in the top part of Fig. 2. It basically predicts that when the concentration of phosphorus $P(t)$ in water at week t is relatively high, then the biomass of algae next week $BIO(t+1)$ will be relatively low and vice versa. Since phosphorus is food for algae, it would seem more likely that this relation would be exactly the opposite.

The discovery of this unexpected pattern in the Retis trees was so surprizing to one of the domain experts that he for a while refused to look further at these results, believing that there was an error in the Retis program. Later the following explanation for the Retis discovery was found (see Fig. 2). Phosphorus was measured in water. In addition to phosphorus in water, there is also some phosphorus in the algae. The total amount of phosphorus in the lagoon is approximately constant. So when there is more algae, there is less phosphorus in water. This further means that when the measured phosphorus is low the biomass of algae must be high and vice versa. As the biomass of algae cannot change completely in one week, high biomass at week t also indicates high biomass at week $t+1$. These relations are formalised in Fig. 2. The

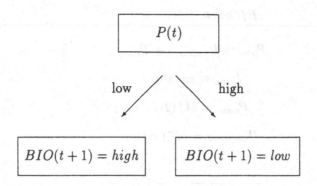

Figure 1: An induced qualitative relation between the concentration of phosphorus this wee and the biomass of algae next week.

bottom of Fig. 2 shows a qualitative version of a regression tree fragment which is then (qualitatively) equivalent to the one in Fig. 1. When this qualitative explanation was found, the doubting expert agreed and suggested that "Maybe this relation always holds between $P(t)$ and $BIO(t + 1)$". Indeed, this same pattern was found in many regression trees at various places in the induced regression trees, and no fragment was found which would indicate the opposite.

This experience showed that induced models, even if they are of no use as (quantitative) predictors, they may still enable an expert to find interesting and surprizing relations. Expert interpretation is in such cases almost entirely qualitative.

3.2 Red deer population dynamics and forest development

Similar experience regarding numerical accuracy and qualitative interpretation was observed in the deer population application [13]. Questions investigated in this study were: How does the degree of browsing of maple and beech trees depend on the the size of the population of reed deer in the region, and on weather conditions like: winter and summer temperatures, periods with snow etc.? How does the weight of the animals depend on their age, weather conditions and the condition of the forest (including degree of browsing of trees)? The goal was to find a policy of forest management and

$$BIO(t) \approx BIO(t+1)$$

$$P_{total} = P_{measured} + P_{algae}$$

$$P_{total} \approx constant$$

$$P_{algae} = M^+(BIO)$$

$$P_{measured} = M^-(BIO)$$

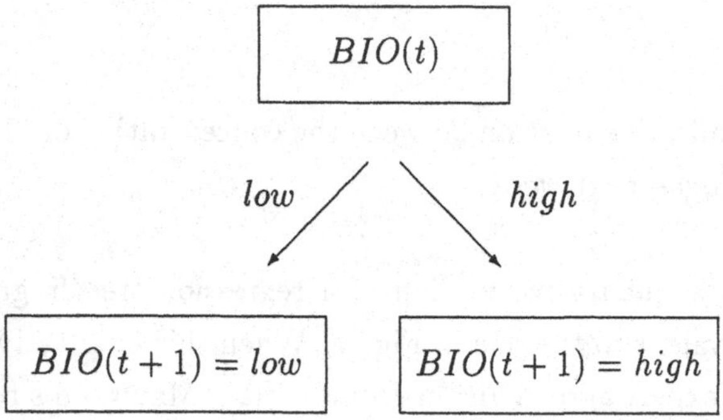

$$\boxed{BIO(t)}$$

low / \ *high*

$$\boxed{BIO(t+1) = low} \qquad \boxed{BIO(t+1) = high}$$

Figure 2: Top: A qualitative explanation of the relation in Fig. 1 (notation $y = M^+(x)$ means that y is a monotonically increasing function of x). The symbol "$\approx$" means "very rough approximation". Bottom: A qualitative relation that is, according to this explanation, equivalent to the one in Fig. 1.

deer harvesting which would enforce a good balance between the deer population and the condition of the forest. The data analysed with the Retis program concern a 400 square km region Snežnik in Slovenia.

In this domain also, the data was too scarce, noisy and incomplete to allow extraction of reliable numerical relations. However, the forestry experts were able to discern several interesting qualitative relations in the induced regression trees. It was found that the degree of browsing of trees does not critically depend only on the size of deer population as previously believed. The newly discovered relations will probably affect the future policy of deer harvesting.

4 Analysis of discrete event simulation results

Discrete event models are used in operations research to study the performance of a system, such as a manufacturing or a servising system. The performance depends on the organisation of the system and the resources allocated to the activities in the system. For example, in a supermarket, the waiting queue in front of the cashiers depends on the number of cashiers. And when potential customers in front of the supermarket decide whether they would enter the supermarket, they will consider the length of the queues in front of the cashiers and decide whether it would be worth waiting. This decision of course also depends on the mentality of the customers, but in any case the chances that a customer will enter are higher when the queues are shorter. An interesting parameter is then the percentage of the potential customers in front of the supermarket who actually decide to enter. A question of practical interest is: How does this percentage of potential customers ("customer utilisation") depend on the number of cashiers and customers arrival rate?

Having a discrete event model of such a system, any particular question can be answered by simulation. That is, for any given number of cashiers, arrival rate and customers waiting tolerance, it is possible by simulation to determine the customer utilisation ratio. An analysis of a system then typically consists of running the simulation for various combinations of the system parameter settings (e.g. varying the number of cashiers and arrival rate etc.), and then tabulating the resulting utilisation rates. The results of such discrete event simulations have the form of tables of numbers. The analyst then has to interpret these tables and find general trends and explicit regularities. As our discrete event simulation expert said: "The task is to find out how the system works." This then enables a decision-maker to reason about re-designing the system or re-allocating the resources in order to effectively improve the performance of the system.

The interpretation of simulation results is known to be a difficult and creative task. Therefore any tools that help in the interpretation of the simulation results tables are of great interest to practitioners in operations research. In [9], the Retis program was applied as such a tool. Our domain expert then tried to interpret the induced regression trees. Very similarly as in the ecological domains, the expert's interpretations were largely qualitative. Of course, here some actual numerical threshold values of the parameters of the system, introduced by Retis, were of interest. When observing the expert's efforts to understand the trees, it occurred to us that those regression trees with some 30 or 40 nodes were actually quite unfriendly and several ideas immediately appeared how the trees could be made more friendly. Nevertheless, the expert rather liked the trees and said that Retis was "the most helpful data intepretation tool of anything that he had used previously".

Later the ILP (Inductive Logic Programming) program Markus [4] was applied to the same data. Since Markus is an ILP program, it can accept domain-specific background knowledge as well as the learning data. For this application, background knowledge simply consisted of arithmetic operations and a numerical comparison predicate. This was sufficient for Markus to discover a very succint theory from the supermarket data. In one experiment the predicate to be induced was:

```
shop90( Cashs, QWsize, MTBArriv)
```

The predicate shop90 is true when the customer utilisation rate is less than 90%, that is more than 10% of potential customers are lost. Here Cashs is the number of cashiers, QWsize is the critical size of the waiting queue at which potential customers stop entering the supermarket, and MTBArriv is the mean time between arrivals. The induced definition of shop90, in Prolog as generated by Markus, but slightly polished for readability, is:

```
shop90( Cashs, QWsize, MTBArriv)  :-
    MTBArriv * Cashs =< 5.

shop90( Cashs, QWsize, MTBArriv)  :-
    MTBArriv =< 3,
    QWsize * Cashs =< 3.

shop90( 1, 1, MTBArriv).
```

This formulation, enabled by the use of (very simple) background knowledge, is much better than regression trees and enables easy planning of the number of cashiers.

This application indicates again the expert's interest in the interpretation which is largely qualitative. He was interested in the question "How the system works?" This study also shows that regression trees are not a very friendly representation. But even so they can be of great interest to the user. A much more explicitly meaningful description was induced by an ILP program. ILP allowed a better representation through the use of domain-relevant background knowledge. The next section describes an application in which the use of background knowledge was absolutely crucial.

5   Predicting mutagenicity

Srinivasan et al. [12] describe an application of ILP to the determination of the mutagenicity of chemical compounds.

The study of Structure/Activity Relationships (SAR) of chemical compounds forms the basis of rational drug design. Thus the determination of SAR is an important area of current research. One widely used method of SAR uses regression to predict activity from attribute-value descriptions of molecules. Such descriptions take into account global properties of molecules, such as hydrophobicity, and molar reflectivity. Regression and many other approaches, including neural networks, are largely limited to such representations. They take into account the global attributes of a molecule, but do not comprehensively consider the *structural relationships* in the molecule.

ILP allows, however, that the complete structural information is taken into account. Srinivasan et al. [12] have applied the *Progol* ILP system [10] to the problem of identifying Ames test mutagenicity within a series of heteroaromatic nitro compounds. They used as examples for learning 230 compounds studied by Hansch and coworkers using classical regression [2]. All the compounds were represented relationally in terms of atoms, bonds and their partial charge. This information was automatically generated by the modelling program QUANTA^{TM}, and was represented as about 18300 Prolog facts for the entire set of 230 compounds.

Srinivasan and coauthors [12] reported that the estimated accuracy of the theory induced by *Progol* was at least as high as the accuracy of both the regression analysis of Hansch and coworkers, and a more recent effort using neural networks. However, on the subset of 42 compounds which had been known as "regression unfriendly", *Progol*'s accuracy was clearly higher than the accuracy of the other approaches.

Perhaps even more important than the predictive accuracy, it should be noted that *Progol*'s theory is easier to comprehend and was generated automatically, without access to any structural indicator variables hand-crafted by experts specifically for this problem. The participating domain expert (M. Sternberg) considered that the most important result of this application was *Progol*'s discovery of a previously unsuspected pattern in molecular structures. *Progol*'s theory provides the new chemical insight that the presence of a five-membered aromatic carbon ring with a nitrogen atom linked by a single bond followed by a double bond indicates mutagenicity. *Progol* has therefore identified a new structural feature that is an alert for mutagenicity.

6   Discussion

Experience from the applications presented in this paper can be summarised as follows. The main interest of the participating experts is in the discovery, from data, of interesting patters that lend themselves to meaningful interpretation. That is, interpretation has to make sense with respect to the expert's prior knowledge ("background knowledge"). Experts are interested in improving their own understanding of the domain of exploration. One expert expressed this by saying: "When I investigate a system, I want to find out *how the system works.*" *Understanding* a system is not the same as having a black box that makes (in a misterious way) correct prediction in any situation. Understanding will allow the expert to reason about changes in the system and, for example, redesign parts of the system or design appropriate control actions. Understanding "How the system works" can be useful even when it is not quite sufficient for making completely accurate predictions.

This expert's interest in understanding and interpretation of induced descriptions is reflected in the following concrete observations in the applications described:

- The expert did not particularly care for predictive accuracy of induced descriptions.

- The expert was not impressed by short prediction rules or small trees, which should be preferred according to the generally accepted principles like MDL (Minimum Description Length) or Occam's rasor. Often it was just the opposite, the expert was insisiting that a bigger tree "tells more".

- Even in cases when induced descriptions were useless as numerical predictors because of their inaccuracy, they were still of interest to experts whenever they offered some sensible qualitative interpretation. Sometimes surprizing qualitative relations may be discoverd.

These observations that continuously recur in ML applications indicate that the induced descriptions are used by experts not so much as predictors, but usually as material for thought. Induced descriptions provide to experts a source of ideas for thinking about their problem domains. They lead to new hypotheses used in further investigation.

The key here then is the *interpretability* of the induced descriptions. The criterion of interpretability is orthogonal to the usual predictive accuracy and related criteria. As interpretability is, similar to comprehensibility, almost completely unexplored, it presents a great challenge for future research in ML. There is little known about how to tackle the notion of interpretability formally. We conclude by some initial thoughts that were inspired by the applications described in this paper. These at least indicate some ingredients of interpretability.

- For a description to be interpretable by an expert, it should connect naturally to the expert's *background knowledge*. That is, a hypothesis should be explainable in terms of background knowledge.

- People like *simplification* and *approximation* when the true relations are complex. Background knowledge helps in such cases to recover the missing details.

- People often also like *redundancy* in a description. Redundancy may be useful for confirmation, alternative "proof", enables flexible use, e.g. more flexibility in measurement, etc.

The above "ingredients of interpretability" are somewhat in contradiction. What is a good balance between simplicity and redundancy is of course an open problem. Anyway, when this problem is explored, it may well turn out that the traditional ML goals of accuracy and statistical significance will sometimes even have to be sacrificed for complexity. Speculating even further, the MDL principle may be preserved, but this time standing for "Maximal Description Length".

References

[1] Bratko, I., Mozetič, I., Lavrač, N. (1989) *KARDIO: A Study in Deep and Qualitative Knowledge for Expert Systems*, MIT Press.

[2] Debnath, A.K., Lopez de Compadre, R.L., Debnath, G., Schusterman, A.J. and Hansch, C. (1991) *Jnl. Medicinal Chemistry*, **34**, 786–797.

[3] Džeroski, S., Grbović, J., Walley, W. (1997) Machine learning applications in biological classification of river water quality. In: *Machine Learning and Data Mining: Methods and Applications* (eds. R.S. Michalski, I. Bratko, M. Kubat) Wiley (in press).

[4] Grobelnik, M. (1992) Markus: an optimized model inference system. *Logic Approaches to Machine Learning Workshop*, Vienna, August 1992.

[5] Karalič, A. (1992) Employing Linear Regression in Regression Tree Leaves. *Proc. 10th European Conference on Artificial Intelligence*, pp. 440-441, Vienna, 1992.

[6] Kompare, B. (1995) *Artificial Intelligence Techniques in Ecological Modelling.* Ph.D. Thesis, Kopenhagen: Royal School of Pharmacy.

[7] Kononenko, I., Bratko, I. (1991) Information based evaluation criterion for classifier's performance. *Machine Learning Journal*, Vol. 6, pp. 283-312.

[8] Križman, V. (1994) Handling noisy data in automated modelling of dynamical systems. M.Sc. Thesis, Faculty of Computer and Info. Sc., University of Ljubljana (in Slovenian).

[9] Mladenić, D., Bratko, I., Paul, R.J., Grobelnik, M., Using machine learning techniques to interpret results from discrete event simulation. *Proc. ECML-94 Conf.*, Catania, Italy, april 1994.

[10] Muggleton, S. (1995) Inverse entailment and Progol. *New Generation Computing*, Vol. 13, Nos. 3,4 (Special issue on Inductive Logic Programming), pp. 245-286.

[11] Pirnat, V., Kononenko, I., Janc, T., Bratko, I. (1989), Medical analysis of automatically induced diagnostic rules. *Third Int. Conf. AI in Medicine*, London.

[12] Srinivasan, A., Muggleton, S.H., King, R.D., Sternberg, M.J.E. (1994) Mutagenesis: ILP experiments in a non-determinate biological domain. In *Proc. Fourth Int. Workshop on Inductive Logic Programming ILP-94*, Bad Honnef/Bonn.

[13] Stankovski, V., Debeljak, M., Bratko, I., Adamič, M. (1996) Modelling the population dynamics of red deer (cervus elaphus L.) with regard to forest development. *Ecological Modelling Conf.*, Copenhagen; to appear in *Ecological Modelling Journal*.

PREPROCESSING BY A COST-SENSITIVE
LITERAL REDUCTION ALGORITHM: REDUCE

N. Lavrac
J. Stefan Institute, Ljubljana, Slovenia

D. Gamberger
R. Boskovic Institute, Zagreb, Croatia

P. Turney
National Research Council Canada, Ottawa, Ontario, Canada

Abstract

This study is concerned with whether it is possible to detect what information contained in the training data and background knowledge is relevant for solving the learning problem, and whether irrelevant information can be eliminated in preprocessing before starting the learning process. A case study of data preprocessing for a hybrid genetic algorithm shows that the elimination of irrelevant features can substantially improve the efficiency of learning. In addition, cost-sensitive feature elimination can be effective for reducing costs of induced hypotheses.

1   Introduction

The problem of relevance was addressed in early research on inductive concept learning [11]. Recently, this problem has also attracted much attention in the context of feature selection in attribute-value learning [1,5,14]. Basically one can say that all learners are concerned with the selection of 'good' literals or features which will be used to construct the hypothesis.

This study is concerned with whether it is possible to detect what information contained in the training data and background knowledge is relevant for solving the learning problem,

---

[1]ISSEK Workshop *Mathematical and Statistical Methods in AI*, September 19–21, 1996, Udine, Italy

and whether irrelevant information can be eliminated in preprocessing before learning. An important difference between our approach and most other approaches is that, when deciding about the relevance of literals, we are concerned with finding 'globally relevant' literals w.r.t. the entire set of training examples, as opposed to finding the 'good literals' in the given local training set (i.e., a set of examples covered by the currently constructed rule in rule induction systems, or a set of covered examples in the current node of a decision tree in TDIDT systems). This is important since the elimination of globally irrelevant literals guarantees that literal elimination will not harm the hypothesis formation process and that during the reduction of the hypothesis space the optimal problem solution will not be discarded. The aim of this study is to distinguish between a set of literals that are relevant for learning and a set of irrelevant literals that can be discarded before learning (i.e., before even entering the 'good literal' competition). Such filtering of irrelevant literals can thus be viewed as a part of preprocessing of the set of training examples.

This paper presents a case study of data preprocessing for a hybrid genetic algorithm which shows that the elimination of irrelevant features can substantially improve the efficiency of learning. In addition, cost-sensitive feature elimination can be effective for reducing costs of induced hypotheses.

The paper is organized as follows: Section 2 introduces the representational formalism, the so-called p/n pairs of examples, gives the definition of irrelevant literals and presents a theorem which is the basis for literal elimination. Section 3 presents the cost-sensitive literal elimination algorithm REDUCE. Section 4 introduces the problem domain, the 20 and the 24 trains East-West Challenges, and presents the results of our experiments that show that the performance of a hybrid genetic algorithm RL-ICET [16] can be significantly improved by applying REDUCE in preprocessing of the dataset. The results of RL-ICET are also compared to those of C4.5 [13].

2   Relevance of literals and features

Consider a two-class learning problem where training set E consists of positive and negative examples of a concept, and examples $e \in E$ are tuples of truth-values of terms in a hypothesis language. The set of all terms, called *literals*, is denoted by L.

2.1   Representation of training examples

Let us represent the training set E as a table where rows correspond to training examples and columns correspond to literals. An element in the table has the value *true* when the example satisfies the condition (literal) in the column of the table, otherwise its value is *false*. If the training set does not have the form of tuples of truth-values, a transformation to this form is performed in preprocessing of the training set.

2.1.1 Learning of propositional descriptions

In the attribute-value learning setting, the transformation procedure is based on analysis of the values of examples in the training set. For each attribute A_i, let v_{ix} ($x = 1..k_{ip}$) be the k_{ip} different values of the attribute that appear in the positive examples and let w_{iy} ($y = 1..k_{in}$) be the k_{in} different values appearing in the negative examples. The transformation results in a set of literals L:

- For discrete attributes A_i, literals of the form $A_i = v_{ix}$ and $A_i \neq w_{iy}$ are generated.

- For continuous attributes A_i, literals of the form $A_i \leq (v_{ix} + w_{iy})/2$ are created for all neighboring value pairs (v_{ix}, w_{iy}), and literals literals $A_i > (v_{ix} + w_{iy})/2$ for all neighboring pairs (w_{iy}, v_{ix}). The motivation is similar to that suggested in [2].

- For integer valued attributes A_i, literals are generated as if A_i were both discrete and continuous, resulting in literals of four different forms: $A_i \leq (v_{ix} + w_{iy})/2$, $A_i > (v_{ix} + w_{iy})/2$, $A_i = v_{ix}$, and $A_i \neq w_{iy}$.

2.1.2 Inductive logic programming

Inductive logic programming (ILP) refers to first-order learning of relational descriptions in the representation formalism of logic programs [7]. In this setting, a LINUS transformation approach is assumed that is appropriate for a limited hypothesis language of constrained nonrecursive clauses [6,7]. For example, if the training examples about the target relation $daughter(A_1, A_2)$ are given in the training set, and the background knowledge consists of the definitions of a unary relation $female$ and binary relation $parent$, the transformation of training examples results in a matrix of binary values $true$ and $false$ whose rows correspond to training examples, and columns correspond to the following literals: $female(A_1)$, $female(A_2)$, $parent(A_1, A_2)$, $parent(A_2, A_1)$, $parent(A_1, A_1)$, $parent(A_2, A_2)$, and $A_1 = A_2$.

2.2 p/n pairs of examples and relevance of literals

Assume the set of training examples E represented by a truth-value table where columns correspond to the set of literals L, and rows are tuples of truth-values of literals, representing training examples e_i. The table is divided in two parts, P and N, where P are the positive examples, and N are the negative examples. We use $P \cup N$ to denote the table E.

To enable a formal discussion of the relevance of literals, the following definitions are introduced:

Definition 1. *A p/n pair is a pair of training examples where $p \in P$ and $n \in N$.*

Definition 2. *Literal $l \in L$ covers a p/n pair if in column l of the table of training examples E the positive example p has value true and the negative example n has value false. The set of all p/n pairs covered by literal l will be denoted by $E(l)$.*

Definition 3. *Literal l covers literal l' if $E(l') \subseteq E(l)$.*

To illustrate the above definitions consider a simple learning problem with 5 training examples: three positive p_1, p_2 and p_3, and two negative n_1 and n_2, described by truth-values of literals $l_i \in L$. The truth-value matrix E, showing just some of the truth-values, is given in Table 1.

Examples				l_2			Literals l_4				l_8		
		...	...		...	...		...	...	...		...	...
P	p_1												
	p_2			true			true				false		
	p_3												
N	n_1			false			false				true		
	n_2			false			true				true		

Table 1: Coverage of literals, coverage of p/n pairs.

Literal l_2 in Table 1 seems to be relevant for the formation of the inductive hypothesis since it is true for a positive example and false for both negative examples. This is due to the fact that l_2 covers a positive example, and does not cover the negative examples, and is thus a reasonable ingredient of the hypothesis that should cover the positive examples and should not cover the negatives examples.

Literal l_2 covers two p/n pairs: $E(l_2) = \{p_2/n_1, p_2/n_2\}$. Literal l_8 is inappropriate for constructing a hypothesis, since it does not cover any p/n pair: $E(l_8) = \emptyset$. Literal l_4 seems to be less relevant than l_2 and more relevant than l_8; it covers only one p/n pair: $E(l_4) = \{p_2/n_1\}$. Literal l_2 covers l_4 and l_8, and literal l_4 covers l_8, since $E(l_8) \subseteq E(l_4) \subseteq E(l_2)$.

Table 1 thus gives the following intuition: the more p/n pairs a literal covers the more relevant it is for hypothesis formation. This may be formalized by the next definition.

Definition 4a. *Literal l' is irrelevant if there exists a literal $l \in L$ such that l covers l' $(E(l') \subseteq E(l))$. In other words, literal l' is irrelevant if it covers a subset of p/n pairs covered by some other literal $l \in L$.*

Assume that literals are assigned costs (for instance, cost can be a measure of complexity - the more complex the literal, the higher its cost). Let $c(l)$ denote the cost of literal $l \in L$. The definition of irrelevance needs to be modified to take into account the cost of the literals.

Definition 4b. *Literal l' is irrelevant if there exists a literal $l \in L$ such that l covers l' $(E(l') \subseteq E(l))$ and the cost of l is lower than the cost of l' $(c(l) \leq c(l'))$.*

Our claim is that irrelevant literals can be eliminated in preprocessing. This claim is based on the following theorem, which assumes that the hypothesis language $\mathcal{L}$ is rich enough to allow for a complete and consistent hypothesis H to be induced from the set of training examples E.[2]

Theorem 1. Assume a training set E and a set of literals L such that a complete and consistent hypothesis H can be found. Let $L' \subseteq L$. A complete and consistent hypothesis H can be found using only literals from the set L' if and only if for each possible p/n pair from the training set E there exists at least one literal $l \in L'$ that covers the p/n pair.

Proof of necessity: Suppose that the negation of the conclusion holds, i.e., that a p/n pair exists that is not covered by any literal $l \in L'$. Then no rule built of literals from L' will be able to distinguish between these two examples. Consequently, a description which is both complete and consistent can not be found.

Proof of sufficiency: Take a positive example p_i. Select from L' the subset of all literals L_i that cover p_i. A constructive proof of sufficiency can now be presented, based on k runs of a covering algorithm, where k is the cardinality of the set of positive examples ($k = |P|$). In the i-th run, the algorithm learns a conjunctive description h_i, $h_i = l_{i,1} \land \ldots \land l_{i,m}$ for all $l_{i,1}, \ldots l_{i,m} \in L_i$ that are true for p_i. Each h_i will thus be *true* for p_i (h_i covers p_i), and *false* for all $n \in N$. After having formed all the k descriptions h_i, a resulting complete and consistent hypothesis can be constructed: $H = h_1 \lor \ldots \lor h_k$. □

The importance of the theorem is manifold. First, it points out that when deciding about the relevance of literals it will be significant to detect which p/n pairs are covered by the literal. Second, the theorem enables us to directly detect useless literals that do not cover any p/n pair. This theorem is the basis of the REDUCE algorithm for literal elimination.

3   Cost-sensitive literal elimination

3.1   Cost-sensitive literal elimination algorithm REDUCE

Algorithm 1 implements the cost-sensitive literal elimination algorithm, initially developed within the ILLM learner [3]. This algorithm is the core of REDUCE [9].

The complexity of Algorithm 1 is $\mathcal{O}(|L|^2 \times |E|)$, where $|L|$ is the number of literals and $|E|$ is the number of examples. This algorithm can be easily transformed into an iterative algorithm that can be used during the process of generation of literals [8].

---

[2]Hypothesis H is complete if it covers all the positive examples $p \in P$. Hypothesis H is consistent if it does not cover any negative example $n \in N$.

Algorithm 1. Cost-sensitive literal elimination

> **Given:** CL – costs of literals in L
> **Input:** P, N – tables of positive and negative examples, L – set of literals
> > $RP \leftarrow P, RN \leftarrow N, RL \leftarrow L$
> > **for** $\forall\, l_i \in RL$ $(i \in [1, |L|])$ **do**
> > > **if** l_i has value *false* for all rows of RP **then**
> > > > eliminate l_i from RL
> > > > eliminate column l_i from RP and RN tables
> > > **if** l_i has value *true* for all rows of RN **then**
> > > > eliminate l_i from RL
> > > > eliminate column l_i from RP and RN tables
> > > **if** l_i is covered by any $l_j \in RL$ for which $c(l_j) \leq c(l_i)$ **then**
> > > > eliminate l_i from RL
> > > > eliminate column l_i from RP and RN tables
> > **endfor**
> **Output:** RP, RN – reduced tables of positive and negative examples, RL – reduced set of
> literals

The algorithm can be efficiently implemented using simple bitstring manipulation on the table of training examples E. For this purpose, the table E is transformed into E_t as follows:

> $\forall p \in P$: replace *true* by 1 and *false* by 0
> $\forall n \in N$: replace *false* by 1 and *true* by 0

In this representation, examples $e \in E_t$ and literals $l \in L$ are bitstrings. Coverage can now be checked by set inclusion. Recall that literal l covers literal l' if $E_t(l') \subseteq E_t(l)$. Thus, if l' has value 1 only in (some of) those rows as l has 1 and in no other rows, l' can be eliminated.

3.2 Relevance of features

The term *feature* is used to denote positive literals such as for example $A_i = v$, $A_j \leq w$, and $r(A_i, A_j)$. In the hypothesis language, the existence of one such feature implies the existence of two complementary literals: a positive and a negative literal. Suppose that we consider the feature $Color = black$ and that the attribute $Color$ has three possible values: *black, white, red*. Since each feature implies the existence of two literals, the necessary and sufficient condition that a feature can be eliminated as irrelevant is that both of its literals $Color = black$ and $Color \neq black$[3] are irrelevant. This statement directly implies the procedure taken in our experiment. First we convert the starting feature vector to the corresponding literal vector which has twice as many elements. After that, we eliminate the irrelevant literals and, in the third step, we construct the reduced set of features which includes all the features which have at least one of their literals in the reduced literal vector.

It must be noted that direct detection of irrelevant features (without conversion to and from the literal form) is not possible except in the trivial case where two (or more) features have identical columns in table E_t. Only in this case a feature f exists whose literals f and $\neg f$ cover both literals g and $\neg g$ of some other feature. In a general case if a literal of feature f

---

[3]We use either the notation $\neg(Color = black)$ or $Color \neq black$ to denote a negative literal.

covers some literal of feature g then the other literal of feature g is not covered by the other literal of feature f. But it can happen that this other literal of feature g is covered by a literal of some other feature h. This means that although there does not exist a feature f that covers both literals of the feature g, feature g can be irrelevant.[4]

4   Utility study: The East-West challenge

Michie et al. [12] issued a "challenge to the international computing community" to discover low size-complexity Prolog programs for classifying trains as Eastbound or Westbound. The challenge was inspired by a problem posed by Michalski and Larson [10].

The original challenge [12] included three separate tasks. Michie later issued a second challenge, involving a fourth task. Our experiments described here involve the first and fourth tasks. The first task was to discover a simple rule for distinguishing 20 trains, 10 Eastbound and 10 Westbound, whereas the fourth task involved 24 trains, 12 Eastbound and 12 Westbound. In these two tasks, the set of trains was classified into East and West using an arbitrary human-generated rule (theory). The challenge was to discover the human-generated theories or a simpler theory (rule).

For both tasks, the winner was decided by representing the rule as a Prolog program and measuring its size-complexity. The size-complexity of the Prolog program was calculated as the sum of the number of clause occurrences, the number of term occurrences, and the number of atom occurrences.

4.1   RL-ICET

A cost-sensitive algorithm ICET was developed for generating low-cost decision trees [15]. ICET is a hybrid of a genetic algorithm and a decision tree induction algorithm. The genetic algorithm is Grefenstette's GENESIS [4] and the decision tree induction algorithm is Quinlan's C4.5 [13]. ICET uses a two-tiered search strategy. On the bottom tier, C4.5 uses a TDIDT (Top Down Induction of Decision Trees) strategy to search through the space of decision trees. On the top tier, GENESIS uses a genetic algorithm to search through the space of biases.

ICET takes feature vectors as input and generates decision trees as output. The algorithm is sensitive to both the cost of features and the cost of classification errors. The East-West Challenge involves data in the form of relations, and theories in the form of Prolog programs. For the East-West Challenge, ICET was extended to handle Prolog input. This algorithm is called RL-ICET (Relational Learning with ICET) [16].

---

[4]This analysis helps us to see that the standard approach to rule construction which is based on feature selection is sub-optimal. Most rule learners use a two-phase approach: first, the best feature is selected, and second, for a selected feature, one of the two complementary literals (positive or negative) is used to construct a rule. We suggest that learning should be based on tuples of truth-values of positive and negative literals, rather than on feature vectors.

RL-ICET is similar to the LINUS learning system [6,7] since it uses a three-part learning strategy. First, a preprocessor translates the Prolog relations and predicates into a feature vector format. The preprocessor in RL-ICET was designed specially for the East-West Challenge, whereas LINUS has a general-purpose preprocessor. Second, an attribute-value learner applies a decision tree induction algorithm (ICET) to the feature vectors. Each feature is assigned a cost, based on the size of the fragment of Prolog code that represents the corresponding predicate or relation. A decision tree that has a low cost corresponds (roughly) to a Prolog program that has a low size-complexity. When it searches for a low cost decision tree, ICET is in effect searching for a low size-complexity Prolog program. Third, a postprocessor translates the decision tree into a Prolog program. Postprocessing with RL-ICET is done manually, whereas LINUS performs post-processing automatically.

4.2   Feature construction in RL-ICET

Much of the success of RL-ICET in the East-West challenge tasks may be attributed to its preprocessor which translates the Prolog descriptions of the trains into a feature vector representation.

The data about each train in the East-West challenge were represented using Prolog. For example, the first train, shown below, is represented by the following Prolog clause:

```
eastbound([c(1, rectangle, short, not_double, none, 2, l(circle,1)),
    c(2, rectangle, long, not_double, none, 3, l(hexagon, 1)),
    c(3, rectangle, short, not_double, peaked, 2, l(triangle, 1)),
    c(4, rectangle, long, not_double, none, 2, l(rectangle, 3))]).
```

The relatively compact Prolog description was converted by a simple Prolog program into a feature vector format (tuples of truth-values of features) to be used for decision tree induction. This resulted in rather large feature vectors of 1199 elements. The large vectors were required to ensure that all the features that are potentially interesting for the final solution are made available for ICET.

What follows is a brief outline of the feature construction procedure that occurs in the preprocessing of the training set. We started with 28 predicates that apply to the cars in a train, such as ellipse(C), which is true when the car C has an elliptical shape. For each of these 28 predicates, we defined a corresponding feature. All of the features were defined for whole trains, rather than single cars, since the problem is to classify trains. The feature ellipse, for example, has the value *true* when a given train has a car with an elliptical shape. Otherwise ellipse has the value *false*. We then defined features by forming all possible unordered pairs of the original 28 predicates. For example, the feature ellipse_triangle_load has the value *true* when a given train has a car with an elliptical shape that is carrying a triangle load, and *false* otherwise. Note that the features ellipse and triangle_load may have the value *true* for a given train while the feature `ellipse_triangle_load` has the value *false*, since `ellipse_triangle_load` only has the value *true* when the train has a car that is both elliptical and carrying a triangle load. Next we defined features by forming all possible ordered pairs of the original 28 predicates, using the relation `infront(T, C1, C2)`. For example, the feature `u_shaped_infront_peaked_roof` has the value *true* when the train has a U-shaped car in front of a car with a peaked roof, and *false* otherwise. Finally, we added 9 more predicates that apply to the train as a whole, such as `train_4`, which has the value *true* when the train has exactly four cars. Thus a train is represented by a feature vector, where every feature has either the value *true* or the value *false*.

Each feature was assigned a cost, based on the complexity of the fragment of Prolog code required to represent the given feature. The complexity of a Prolog program is defined as a sum of the number of clause occurrences, the number of term occurrences and the number of atom occurrences. Table 2 shows some examples of constructed features and their costs.

Feature	Prolog Fragment	Cost (Complexity)
ellipse	$has\_car(T,C), ellipse(C)$.	5
short_closed	$has\_car(T,C), short(C),$ $closed(C)$.	7
train_4	$len1(T,4)$.	3
train_hexagon	$has\_load1(T, hexagon)$.	3
ellipse_peaked_roof	$has\_car(T,C), ellipse(C),$ $arg(5, C, peaked)$.	9
u_shaped_no_load	$has\_car(T,C), u\_shaped(C),$ $has\_load(C,0)$.	8
rectangle_load_infront *_jagged_roof*	$infront(T, C1, C2),$ $has\_load0(C1, rectangle),$ $arg(5, C2, jagged)$.	11

Table 2: Examples of features and their costs.

The feature vector for a train does not capture all the information that is in the original Prolog representation. For example, we could also define features by combining all possible unordered triples of the 28 predicates. However, these features would likely be less useful, since they are so specific that they will only rarely have the value *true*. If the target concept should happen to be a triple of predicates, it could be closely approximated by the conjunction of the three pairs of predicates that are subsets of the triple.[5]

4.3   Previous results of RL-ICET

RL-ICET was the winning algorithm for the second task in the first East-West Challenge, and it performed very well in the other three tasks.

---

[5]This kind of translation to feature vector representation could be applied to many other types of structured objects. For example, consider the problem of classifying a set of documents. The keywords in a document are analogous to the cars in a train. The distance between keywords or the order of keywords in a document may be useful when classifying the document, just as the *infront* relation may be useful when classifying trains.

Since the initial population of biases in ICET is set randomly, ICET may produce a different result each time it runs. Therefore, when solving a problem, ICET needs to be run several times. The best (lowest cost) decision tree that was generated for the first competition [16] is shown below. The total cost of the tree equals 18 units (obtained as a sum of short_closed = 7, train_4 = 3, u_shaped = 5, and train_circle = 3).

```
short_closed = 1: 1 (8.0)
short_closed = 0:
|    train_4 = 0: 0 (7.0)
|    train_4 = 1:
|    |   u_shaped = 0: 0 (2.0)
|    |   u_shaped = 1:
|    |   |    train_circle = 0: 0 (1.0)
|    |   |    train_circle = 1: 1 (2.0)
```

The induced decision trees were converted into Prolog programs by hand. For example, the above decision tree was converted to the following Prolog program.

```
eastbound(T) :-
    has_car(T, C),
    ((short(C), closed(C));
    (len1(T, 4), u_shaped(C), has_load1(T, circle))).
```

The above Prolog program was the entry for the first competition. This program has a complexity of 19 units, which shows that the cost of the decision tree (18 units) is only an approximation of the cost of the corresponding Prolog program, since some Prolog code needs to be added to assemble the Prolog fragments into a working whole. This extra code means that the sum of the sizes of the fragments is less than the size of the whole program. It is also sometimes possible to subtract some code from the whole, because there may be some overlap in the code in the fragments. The ideal solution to this problem would be to add a post-processing module to RL-ICET that automatically converts the decision trees into Prolog programs. The complexity could then be calculated directly from the output Prolog program, instead of the decision tree. Although post-processing with RL-ICET was done manually, it could be automated, as demonstrated by LINUS, which has a general-purpose post-processor.

4.4   Feature elimination by REDUCE

The objective of the experiments was to show the utility of the literal elimination algorithm REDUCE. Two experiments were performed separately for the 20 and 24 trains problems. In both experiments, the RL-ICET preprocessor was used to generate the appropriate features and to transform the training examples into a feature vector format. This resulted in two training sets of 20 and 24 examples each, described by 1199 features.

In order to apply the REDUCE algorithm we first converted the starting feature vector of 1199 elements to the corresponding literal vector which has twice as many elements, containing 1199 features generated by the RL-ICET preprocessor (positive literals) as well as their negated counterparts (1199 negative literals). After that, we eliminated the irrelevant literals and, in the third phase, constructed the reduced set of features which includes all the features which have at least one of their literals in the reduced literal set.

The experimental setup, designed to test the utility of REDUCE, was as follows. First, 10 runs of the ICET algorithm were performed on the set of training examples described with 1199 features. Second, 10 runs of ICET were performed on the training examples described with the reduced set of features selected by REDUCE.

Ten runs were needed because of the stochastic nature of the ICET algorithm: each time it runs, it yields a different result (assuming that the random number seed is changed). If we compared one single run of ICET on 1199 features to one run of ICET on the reduced feature set, the outcome of the comparison could be due to chance.

The results were compared with respect to execution times, costs of decision trees induced by ICET, and the complexity of Prolog programs after the RL-ICET transformation of decision trees into the Prolog program form (notice that the transformation into the Prolog form is currently manual and sub-optimal, which means that a tree with lowest cost found by ICET is not necessarily transformed into a Prolog program with lowest complexity).

The results of the experiment are summarized in Tables 3 and 4. The average results of 10 runs of RL-ICET were compared with respect to the costs of decision trees and execution times. Notice that all the experiments are independent of each other, e.g., results of experiment 4 should not be compared to the results of experiment 14. Only average results are relevant for the comparison.

4.4.1   Results of the 20 trains experiment

With the 20 train data, REDUCE cut the original set of 1199 features down to 86 features. In this way, the complexity of the learning problem was reduced to about 7% (86/1199) of the initial learning problem. Results of 10 runs of ICET on the 1199 feature set are the results reported in [16], whereas results of 10 runs of ICET on the training examples described with 86 features are new.

The results show that the efficiency of learning significantly increased. In the initial problem with 1199 features, the average time per experiment was about 2 hours and 17 minutes, whereas in the reduced problem setting with 86 features the average time per experiment was about 12 minutes. The difference between times t_1 and t_2 is significant at the 99.99% confidence level. This shows the utility of literal reduction for genetic algorithms which are typically expensive in terms of CPU time.

The average cost of descriptions induced from the 86 feature set has decreased (from 20 to 18.6), but the difference between decision tree costs c_1 and c_2 is not significant. The variance (or the standard deviation) of the costs was also reduced, i.e., the costs of the decision trees

| 86 $features$ | | | | 1199 $features$ | | |
$Trial$	$Time$ t_1	$Cost$ c_1	$Compl.$ cm_1	$Trial$	$Time$ t_2	$Cost$ c_2	$Compl.$ cm_2
1	11 : 05	18	22	11	2 : 21 : 32	24	25
2	11 : 19	21	27	12	2 : 21 : 34	21	22
3	12 : 55	18	22	13	2 : 19 : 15	20	22
4	11 : 35	18	22	14	2 : 19 : 32	20	22
5	15 : 16	18	22	15	2 : 16 : 20	18	19
6	11 : 35	18	19	16	2 : 23 : 52	22	23
7	11 : 32	18	22	17	2 : 24 : 09	21	22
8	11 : 38	18	22	18	2 : 18 : 41	16	20
9	11 : 28	18	22	19	2 : 16 : 58	18	22
10	11 : 18	21	23	20	2 : 23 : 09	20	22
Sum	119 : 41	186	223	Sum	23 : 25 : 02	200	219
$Mean$	11 : 57	18.6	22.3	$Mean$	2 : 16 : 54	20	21.9

Table 3: Summary of results in the 20 trains East-West challenge.

generated from 1199 features vary more than the costs of the trees generated from 86 features: $var(c_1) = 1.6$ $(sd(c_1) = 1.3)$ and $var(c_2) = 5.1$ $(sd(c_2) = 2.3)$.

The lowest cost decision tree (16 units) reported in Table 3 was generated in trial 18, with the full set of 1199 features. However, all of the features that appear in the tree in trial 18 are also in the reduced set of 86 features, so REDUCE does not prevent RL-ICET from possibly discovering this tree (if the genetic search algorithm is lucky). Notice that the sub-optimal transformation into the Prolog form transforms the lowest cost decision tree into a Prolog program with complexity 20, which is higher than the minimal size-complexity non-recursive Prolog program (size 19) for the 20 train problem, generated from decision trees generated in trials 6 and 15.

It is important to note that, using the reduced set of 86 features, in Trial 6, the same best tree as reported in [16] and obtained by Trial 15 from 1199 features, was induced (cost = 18, complexity = 19, see Section 4.3). The fact that the same optimal non-recursive Prolog program was induced and the substantial efficiency increase confirm the usefulness of our approach for learning with genetic algorithms.

It is also interesting to observe that the trees induced by RL-ICET from the set of 1199 features use mostly the features of the reduced 86 feature set; however, some other features are used as well (in Trial 11: triangle_load_one_load, in Trials 13, 14 and 20: no_roof_infront_short, in Trial 17: triangle_load).

4.4.2   Results of the 24 trains experiment

In this experiment, REDUCE decreased the number of features from 1199 to 116. In this way, the complexity of the learning problem was reduced to about 10% (116/1199) of the initial learning problem. The results show that the efficiency of learning significantly increased. In the initial problem with 1199 features, the average time per experiment was nearly two hours, whereas in the reduced problem setting with 116 features the average time per experiment was about 14 minutes. The difference between times t_1 and t_2 is significant at the 99.99% confidence level.

116 features				1199 features			
Trial	Time t_1	Cost c_1	Compl. cm_1	Trial	Time t_2	Cost c_2	Compl. cm_2
1	14 : 35	20	33	11	1 : 54 : 15	27	33
2	14 : 26	18	30	12	1 : 55 : 29	21	28
3	14 : 59	18	30	13	2 : 00 : 25	26	31
4	14 : 17	21	34	14	1 : 56 : 31	25	28
5	13 : 32	18	28	15	1 : 56 : 47	25	30
6	13 : 31	22	27	16	1 : 57 : 14	24	27
7	14 : 29	18	28	17	1 : 56 : 52	28	31
8	13 : 54	23	28	18	1 : 56 : 33	23	28
9	13 : 51	23	33	19	1 : 49 : 08	27	30
10	14 : 30	18	29	20	1 : 47 : 46	28	41
Sum	2 : 22 : 04	199	300	Sum	19 : 11 : 00	254	307
Mean	14 : 12	19.9	30	Mean	1 : 55 : 05	25.4	30.7

Table 4: Summary of results in 24 trains East-West challenge.

The average cost of decision trees induced from the 116 feature set has also decreased. The difference between decision tree costs c_1 and c_2 is significant at the 99.99% confidence level. Our hypothesis that variance (standard deviation) of the output of RL-ICET can be reduced is only weakly supported since the inequality of variance is insignificant: $var(c_1) = 4.8$ $(sd(c_1) = 2.2)$ and $var(c_2) = 5.2$ $(sd(c_2) = 2.3)$.

Turney's original entry in Michie's 24 train challenge was the following tree whose total cost equals 23 units:

```
peaked_roof = 1: 1 (6.0)
peaked_roof = 0:
|    double = 1:
|    |    train_diamond = 0: 1 (5.0)
|    |    train_diamond = 1: 0 (1.0)
|    double = 0:
|    |    closed_hexagon_load = 1: 1 (1.0)
|    |    closed_hexagon_load = 0: 0 (11.0)
```

The tree is transcribed into a Prolog program of size 23:

```
eastbound(T) :-
    has_car(T, C),
    (arg(5, C, peaked);
    (double(C), not has_load0(C, diamond));
    (has_load0(C, hexagon), closed(C))).
```

Notice that the cost of this decision tree is only slightly below average for 1199 features, is above average for 116 features, and is higher than the cost of the minimal decision tree induced from the 116 feature set (minimal cost is 18). However, this Prolog program has a lower complexity than any of the 20 Prolog programs generated in this experiment. This tree/Prolog program was generated with all of the _infront_ features disabled.[6] When the _infront_ features are removed, the number of features drops from 1199 to 415.

When comparing the features of the above best tree with the set of 116 features, all features appearing in the tree appear also in the reduced 116 feature set, except closed_hexagon_load. However, the feature closed_infront_closed substitutes for closed_hexagon_load in the 116 feature set. This means that the reduction algorithm could have removed from the feature set either closed_hexagon_load or closed_infront_closed, but has, due to its ignorance about the transformation procedure into Prolog encodings, randomly decided to eliminate closed_hexagon_load. The resulting substition does not change the cost of the tree (cost = 23), but the corresponding Prolog program is larger (complexity = 26):

```
peaked_roof = 1: 1 (6.0)
peaked_roof = 0:
|    double = 1:
|    |    train_diamond = 0: 1 (5.0)
|    |    train_diamond = 1: 0 (1.0)
|    double = 0:
|    |    closed_infront_closed = 1: 1 (1.0)
|    |    closed_infront_closed = 0: 0 (11.0)

eastbound(T) :-
    has_car(T, C),
    (arg(5, C, peaked);
    (double(C), not has_load0(C, diamond));
    (infront(T, C1, C2), closed(C1), closed(C2))).
```

4.5 Costs versus complexity: Results for 20 and 24 trains

Due to the imperfect transformation of decision trees into Prolog rules, it is hard to achieve the optimal result in terms of the complexity of induced Prolog rules (the goal of the competition

---

[6]Disabling this feature was due to the suspicion that the "infront" features weren't very useful, since S. Muggleton had decided to redefine the "append" predicate by adding cut to it.

was to minimize Prolog code complexity); that is, RL-ICET optimizes the cost of decision trees and not the complexity of Prolog encodings.

To test the correlation between the costs of decision trees and the complexity of Prolog rules we have tried to find the correlation between costs and complexity.

To this end, we have computed the correlation between cost and complexity for the 20 trains experiment: $corr(c_1, cm_1) = 0.73$ and $corr(c_2, cm_2) = 0.86$, and for the 24 trains experiment: $corr(c_1, cm_1) = 0.22$ and $corr(c_2, cm_2) = 0.66$. From these correlations we speculate that:

- the correlation decreases as the number of trains increases, and

- the correlation decreases as the number of features decreases.

Similar conclusions hold also if the correlations between cost and complexity are computed on vectors of 20 elements (and not 10 elements as above), merging the results of the 20 and the 24 trains experiments.

The complexity of the Prolog programs is only weakly correlated with the cost of the decision trees. The ideal correlation (1.0) was not achieved due to the imperfect transformation of decision trees into Prolog rules. This transformation is performed in two steps. In the first step, a Prolog program is created whose structure is nearly identical to the structure of the decision tree. The second step employs ways to compress the Prolog program, by removing redundant sections of code and altering the structure of the program; this step actually disturbs the correlation between the cost of the decision trees and the complexity of the Prolog programs. If the second step were eliminated, there would be a much higher correlation between the cost and the complexity, but there would also be a large increase in the complexity of Prolog programs.

The ideal solution to this problem would be to fully automate the transformation of the decision trees to Prolog programs and then modify RL-ICET to search for the least complex Prolog program, instead of the least costly decision tree.

4.6   Applying C4.5 in the East-West Challenge

We have compared the results of RL-ICET with the ones achieved using C4.5 [13]. This experiment was made in order to check whether our claims of the usefulness of feature reduction can be made more general.

To do so, C4.5 was first applied to the 24 trains problem, using the default settings. C4.5 generated a tree that makes one error on the training data; it misclassifies one of the 24 trains. This is because of one of the default C4.5 parameters settings: the parameter that sets the minimum number of objects that can be at one leaf in the tree. The default value is two. In the experiments on 20 trains and 24 trains, C4.5 was therefore run with the parameter setting which sets the minimum number of objects at one. In this way, C4.5 generates trees that make no errors on the training data.

4.6.1 Results of C4.5

Results of using C4.5 are given in Table 5. Feature reduction does not help C4.5 to find a better solution in terms of costs: using feature reduction, C4.5 becomes slower and gains nothing, since feature reduction itself takes about 5 minutes (real 4:49.00, user 3:32.50, system 0:40.34, times measured on a HP Workstation).

	20 *Trains* 86 *features*	20 *trains* 1199 *features*	24 *trains* 116 *features*	24 *trains* 1199 *features*
Cost	22	22	23	23
Time (seconds)	1	1	1	2

Table 5: Results of C4.5.

4.6.2 Comparing RL-ICET and C4.5

From the comparison of Table 4 and Table 5 we see that for the 116 feature dataset RL-ICET has lower average cost (19.9) than C4.5 (cost 23), but not for the 1199 feature data (average cost 25.4 for RL-ICET and cost 23 for C4.5). From the comparison of Tables 3 and 5 we see that RL-ICET has lower average cost than C4.5 both for the 86 feature dataset (average cost of 18.6 units for RL-ICET and 22 units for C4.5) and for the 1199 feature dataset (average cost of 20 units for RL-ICET and 22 units for C4.5).

In order to see how much literal reduction contributes to the favourable results achieved by RL-ICET, let us separately analyse the results for the reduced and original datasets. For reduced feature sets, the results are favourable for RL-ICET when compared to C4.5.

For the reduced datasets the results are as follows:

86 features, 20 trains: both the minimal cost (18) and the average cost (18.6) are lower than the cost of the tree induced by C4.5 (22).

116 features, 24 trains: both the minimal cost (18) and the average cost (20) are lower than the cost of the tree induced by C4.5 (23).

For the original 1199 feature datasets the results are as follows:

1199 features, 20 trains: both the minimal cost (16) and the average cost (20) of the trees induced by RL-ICET are better than the cost of the tree induced by C4.5 (22).

1199 features, 24 trains: the minimal cost (21) of the tree induced by RL-ICET is lower than the cost of the tree induced by C4.5 (23). However, the average cost of trees induced by RL-ICET (25.4) is higher.

In summary, we may claim that feature reduction helped RL-ICET to achieve very favourable results. In both experiments (20 and 24 trains) it helped RL-ICET to outperform C4.5, when comparing costs of decision trees, both in terms of minimal and average costs. In both experiments it helped RL-ICET to substantially reduce the execution time; however, even on reduced feature sets, RL-ICET of course needs much more time than C4.5.

5 Summary

This work is a study of the problem of relevance for inductive learning system, applicable both in attribute-value (feature vector) learning and in the LINUS transformation approach to inductive logic programming. Irrelevant literal elimination, such as performed by REDUCE, assumes that the goal of the learning algorithm is to find a simple (low size or low cost) hypothesis. This assumption applies to most existing learning systems.

The presented case study of data preprocessing shows that cost-sensitive elimination of irrelevant features can substantially improve the efficiency of learning and can reduce the costs of induced hypotheses. This study, using the hybrid genetic decision tree induction algorithm RL-ICET on two East-West Challenge problems, together with other presented results in feature reduction confirm the usefulness of feature reduction in preprocessing.

In order to evaluate the effects of feature reduction, we have also compared the results of ICET (with and without feature reduction) with the results achieved using C4.5 [13]. In both experiments, feature reduction (reduction to 86 and 116 features, respectively) helped ICET to outperform C4.5 when comparing costs of decision trees, both in terms of minimal and average costs. On the other hand, the application of REDUCE did not help C4.5 itself to induce a lower cost solution from examples described with fewer features.

More detailed information on this study is available in two technical reports of J. Stefan Institute, Ljubljana, that can be obtained upon request from the authors.

Acknowledgements

This research was supported by the Slovenian Ministry of Science and Technology, the Croatian Ministry of Science, the National Reseach Council of Canada and the ESPRIT IV project 20237 Inductive Logic Programming 2. The authors are grateful to Donald Michie for his stimulative interest in this work, and to Sašo Džeroski for his involvement in earlier experiments.

References

1. Caruana, R. and D. Freitag: Greedy Attribute Selection, in: Proceedings of the 11th International Conference on Machine Learning, Morgan Kaufmann, 1994, 28–36.

2. Fayyad, U.M. and K.B. Irani: On the handling of continuous-valued attributes in decision tree generation, Machine Learning, 8 (1992), 87–102.

3. Gamberger, D.: A Minimization Approach to Propositional Inductive Learning, in: Proceedings of the 8th European Conference on Machine Learning, Springer, 1995, 151–160.

4. Grefenstette, J.J.: Optimization of control parameters for genetic algorithms, IEEE Transactions on Systems, Man, and Cybernetics, 16 (1986), 122–128.

5. John, G.H., R. Kohavi and K. Pfleger: Irrelevant Features and the Subset Selection Problem, in: Proceedings of the 11th International Conference on Machine Learning, Morgan Kaufmann, 1994, 190–198.

6. Lavrač, N., S. Džeroski and M. Grobelnik:. Learning Nonrecursive Definitions of Relations with LINUS, in: Proceedings of the 5th European Working Session on Learning, Springer, 1991, 265–281.

7. Lavrač, N. and S. Džeroski: Inductive Logic Programming: Techniques and Applications, Ellis Horwood, 1994.

8. Lavrač, N., D. Gamberger and S. Džeroski: An Approach to Dimensionality Reduction in Learning from Deductive Databases, in: Proceedings of the 5th International Workshop on Inductive Logic Programming, Scientific Report, Katholieke Universiteit Leuven, 1995, 337–354.

9. Lavrač, N., D. Gamberger and P. Turney: Cost-Sensitive Feature Reduction Applied to a Hybrid Genetic Algorithm, in: Proceedings of the 7th International Workshop on Algorithmic Learning Theory, Springer, 1996, 127–134.

10. Michalski, R.S. and J.B. Larson: Inductive Inference of VL Decision Rules, ACM SIGART Newsletter, 63 (1977), 38–44.

11. Michalski, R.S.: A Theory and Methodology of Inductive Learning, in: Machine Learning: An Artificial Intelligence Approach (Eds. R. Michalski, J. Carbonell and T. Mitchell), Tioga, 1983, 83–134.

12. Michie, D., S. Muggleton, D. Page and A. Srinivasan: To the International Computing Community: A new East-West Challenge. Oxford University Computing Laboratory, Oxford, 1994. [Available at URL ftp://ftp.comlab.ox.ac.uk/pub/Packages/ILP/trains.tar.Z.]

13. Quinlan, J.R.: C4.5: Programs for Machine Learning, Morgan Kaufmann, 1993.

14. Skalak, D.: Prototype and Feature Selection by Sampling and Random Mutation Hill Climbing Algorithms, in: Proceedings of the 11th International Conference on Machine Learning, Morgan Kaufmann, 1994, 293–301.

15. Turney, P.: Cost-Sensitive Classification: Empirical Evaluation of a Hybrid Genetic Decision Tree Induction Algorithm, Journal of Artificial Intelligence Research, 2 (1995), 369–409. [Available at URL http://www.cs.washington.edu/research/ jair/home.html.]

16. Turney, P.: Low Size-Complexity Inductive Logic Programming: The East-West Challenge as a Problem in Cost-Sensitive Classification, in: Advances in Inductive Logic Programming (Ed. L. De Raedt), IOS Press, 1996, 308–321.

A GENERAL FRAMEWORK FOR SUPPORTING RELATIONAL CONCEPT LEARNING

L. Saitta
University of Turin, Turin, Italy

ABSTRACT

This paper describes a general representation framework that offers a unifying platform for a number of systems learning concepts in First Order Logic. The main aspects of this framework are discussed, specifically, the separation between the hypothesis logical language and the representation of data by means of a relational database, and the introduction of a functional layer between data and hypotheses, which makes the data accessible by the logical level through a set of abstract properties.

A novelty, in the hypothesis representation language, is the introduction of the construct of internal disjunction; such a construct, first used by the AQ and Induce systems, is here made operational via a set of algorithms, capable to learn it, for both the discrete and the continuous-valued attributes case. These algorithms are embedded in learning systems using different paradigms, such as symbolic, genetic or connectionist ones.

1. INTRODUCTION

Learning knowledge expressed in First Order Logic (FOL) has been an appealing task since the beginning of Machine Learning. Early attempts [1-4] have proposed a number of conceptually interesting ideas, which provided foundations and suggestions for later work. However, the need of excessive computational resources made the proposed methods rather impractical.

However, some systems, such as ML-SMART [5,6] and FOCL [7], have shown that FOL concept learning is not only abstractly interesting, but feasible. The system ML-SMART, in particular, has been applied to a number of real world problems [6,8], and suggested effective solutions for many of the encountered problems. More precisely, the system had special mechanisms to handle noise and continuous attributes, it learned a structured knowledge base (not just a set of flat "condition-action" rules), and it allowed for evidential reasoning. An important aspect of ML-SMART was its interface to a database, from which the examples were extracted, and in which both the generated hypotheses and their extensions were stored [5]. This feature proved to be essential in learning an industrial troubleshooter [8] used in field. ML-SMART evolved later into two new systems, both including inductive and deductive components: SMART+ [9], which emphasizes handling noise and continuous attributes by exploiting powerful heuristics for controlling the search in the hypothesis space, and WHY [10, 11], which stresses the importance of exploiting background knowledge, in the form of a causal model of the application domain, and of learning comprehensible and justifiable knowledge. Moreover, both systems have an embedded machinery, based on relational algebra, for interacting with a relational database.

In the last years, learning in FOL has been re-proposed as "Learning Relations" [12] or Inductive Logic Programming (ILP) [13]. Originally, learning relations was an extension to FOL of the top-down construction of a decision tree, similar to the method presented in [14]. Novelties, with respect to previous FOL learning approaches, were, on one hand, the possibility for the learned concept to contain variables, i.e., to be the name of a relation, an intensional definition of which was to be found from its extensional representation, and, on the other hand, the possibility of learning recursive concepts.

Whatever the name under which learning in FOL is performed, a major issue to be tamed is computational complexity. The matching problem has been proved to be an NP-complete problem [15] even in very simple cases. There are basically two ways to deal with this problem: the first is to reduce the search by adding various kinds of bias in the learning process, both declarative (syntactic and semantic [16]) and procedural ones. The second one is to increase the search power of the learning algorithm. Constraining the search by adding a strong declarative bias, aimed at reducing the expressive power of the hypothesis language, is the solution most frequently adopted in ILP [16]. Systems like SMART+, instead, have a much weaker declarative bias, but they widely exploits procedural biases in the form of search heuristics [9]. Increasing the search power has been tried for the first time (in FOL) in the system REGAL, which learns relations via a genetic algorithm [17, 18].

By its very nature, learning in FOL has traditionally been cast inside the symbolic paradigm. However, the possibility of learning or revising relational knowledge by means of other paradigms, such as the genetic or the connectionist one, could greatly enlarge the class of solvable problems.

A most fundamental issue of learning in FOL is knowledge representation. In particular, both discrete and continuous-valued descriptors have to be handled in a homogeneous way (at least from the user's point of view). Moreover, in view of the application on large databases, a learning system must be able to transparently, quickly and effectively interact with (commercial) database management systems. In this paper, we will describe a knowledge representation framework that shows the above features. A more extensive treatment of the topic can be found in [22]. This framework has several strong points. The first one is the ability to deal with formulas containing the logical construct of internal disjunction. The importance of such a construct, in order to obtain compact and readable knowledge, has been underlined by Michalski since his first works [3, 21], in which he introduced the VL_1, VL_{21} languages. We have developed will present efficient algorithms for learning internal disjunctions, which are embedded in a variety of systems based on different paradigms.

Another aspect to be stressed is the decoupling between the logical level of hypothesis representation and the relational database format in which the examples are stored. The link between the two levels is supplied by a mapping, containing a set of semantic functions, which acts as an interface between the database and the learner. In this way, the learner gives to the interface a formula to verify and receives back the answer, ignoring how the data are represented. This solution has the advantage that performing data mining on a database does not require rewriting the examples in an ad-hoc format (e.g., ground clauses). On the other hand, such a separation between hypotheses and data allows different data representations to be used, by simply changing the semantic mapping.

Finally, the above two features, namely the availability of the internal disjunction and the transparency of data representation achieved through the semantic mapping, allow both discrete and continuous attributes to be dealt with in a uniform way by the learning algorithms.

2. THE KNOWLEDGE REPRESENTATION FRAMEWORK

Many learning algorithms in FOL describe the learning instances using a subset of the hypothesis description language in ground form. This was the method adopted, for instance, by Induce [3, 21], and, more recently, in ILP systems [13]. This method has the advantage that the truth of a formula can be proved using logical resolution only. In the framework we present, on the contrary, we keep separate the hypothesis description language (HDL) and the instance description language (IDL). The HDL consists of a declarative definition of the logical language syntax used for expressing hypotheses. The IDL, instead, includes the capability of representing in a declarative way both the internal *structure* of the examples and their *properties*. In other words, the IDL allows an instance to be represented not only as a compound object, but also in terms of a collection of properties. Then, a complete representation of an instance consists of the specification of its elementary component objects, plus a set of *properties* such as "color of an object", "area of an object", or "relative position (follows/precedes) between two objects", which, together, describe the *structure* of the instance; this representation is close to the style adopted in object oriented databases.

The ability to handle property definitions in the IDL is one of the key notions to understand where the peculiar advantages of the proposed representation come from. Also, this is a major point where IDL differs from just a structural representation of the instances by means of a list of ground predicates. Hypotheses are expressed in a logical language, whose basic predicate semantics is defined in terms of the instance properties specified in the IDL. This representation framework allows different reasoning schemes to be used at the hypothesis and at the instance levels, simplifying the integration of different learning algorithms; moreover, its modularity makes it easy to interface a learning system (working at the hypothesis level) to a database without any preprocessing of the dataset.

2.1 The Instance Description Language (IDL)

When learning propositional knowledge, it is common practice to describe the learning instances as vectors of attributes. We will now extend this framework in order to deal with structured learning instances (or examples) and their properties. Before describing the structural representation of the examples, we need to define the meaning of *concept instance* and of *learning instance* (or *scenario*) in our framework. Given a *concept*, i.e. a name of a relation, a *concept instance* is a group of objects, that, as a whole, can be labelled with that name. More formally, a concept is a n-ary predicate $c(x_1, ... , x_n)$, and a concept instance is an n-tuple $<a_1, ... , a_n>$ of objects such that $c(a_1, ... , a_n)$ is true. A *scenario* is simply a collection of atomic objects. As an analogy, we may think of a scenario as a segmented picture. A scenario may contain several instances of the same concept or even of different concepts. The main characteristic of a scenario is that it constitutes a separate universe, in the sense that the variables occurring in a formula can only be bound to objects occurring in the same scenario.

The logical architecture of the dataset has then three levels of abstraction: scenarios, concept instances, and elementary objects. Each elementary object (simply called "object" in the following) has a vector of attributes associated to it. The restriction on the bounds inside a scenario strongly limits the combinatorial explosion of the size of relations (extensions) associated to hypotheses. According to this view, the best would be to generate one scenario for each concept instance. But, in this case, relations among different instances, such as those involved in recursive definitions, could not be found. Another possibility would be to adopt the opposite solution, i.e., to consider a unique scenario, containing all the instances, and to limit the search by imposing syntactic restrictions on the occurrence of variables. The same scheme easily deals with the problem of representing multiple target concepts.

Let us now consider the definition of object properties. Available domain knowledge could be exploited for this task. Properties are defined using a functional declaration, as in object oriented programming. Let x be an object description. A valid property is a boolean, discrete, or continuous function having as domain the cartesian product X^n, being X the domain of the elementary objects composing the instances. From the implementation point of view, each property definition is a computational expression on a set of predefined functions providing access to the object attributes. It may happen that a property cannot be expressed in analytical form in terms of existing object attributes; in this case, the property is directly represented in extensional form in the database by means of a corresponding relation.

2.2 The Hypothesis Description Language (HDL)

As for IDL, we will first describe the Hypothesis Description Language from an abstract point of view, and, then, we will show how a formula of HDL can be evaluated, in an efficient way, on the adopted learning instance representation by applying a sequence of *relational algebra* operators. The HDL is a clausal language in annotated predicate calculus, equivalent to the VL_{21} language [21], without numerical quantifiers but with negation on atoms. Moreover, the HDL we used allows predicates to contain one internal disjunction. Even though a more complex version of this language, including numerical quantifiers and unrestricted negation has been used in SMART+ [9] and Enigma [8], we will consider here this simpler form that allows a direct comparison with the languages used in other systems such as FOIL, FOCL and most ILP programs.

The reasons for choosing an annotated predicate calculus as HDL, instead of classical Horn clauses, are the compact and readable hypothesis representation (thanks to the internal disjunction construct), the uniform representation of predicates dealing with discrete or continuous attributes, and the easiness in evaluating the predicates on the learning instances (represented in IDL). In the following, after introducing the annotated predicate calculus, each of the above reasons is discussed in detail.

According to the definition given by Michalski [3, 21], in annotated predicate calculus a term occurring in a predicate can be a constant, a variable, or a disjunction of constants. The atomic formula

$$\text{Color}(x, \text{White} \lor \text{Black} \lor \text{Yellow}) \tag{1}$$

is well formed in annotated predicate calculus. As it is intuitive, the meaning of (1) is "the color of object x is either white, or black, or yellow". It is also immediate to verify that formula (1) is equivalent to the disjunction:

$$\text{Color}(x, \text{White}) \lor \text{Color}(x, \text{Black}) \lor \text{Color}(x, \text{Yellow}).$$

In general, every clause in annotated predicate calculus can be syntactically rewritten into an equivalent set of classical clauses, preserving its semantics. Predicates in HDL are of two kinds: *constraint predicates,* which do not contain the internal disjunction term, and *learnable predicates,* containing an internal disjunction. The latter derive their name from the fact that their internal disjunction term can be modified during the learning process. Internal disjunctions are essentially sets of constants of the same type: symbolic (discrete) or real (continuous).

For the sake of notation compactness, we denote finite sets of discrete or symbolic values by $[v_1,...,v_n]$, and infinite sets, corresponding to intervals along unidimensional real space, by $<v_1, v_2>$. Predicates may have any arity, but in many applications it turned out that binary ones are sufficiently expressive. Finally, an *inductive hypothesis* is a formula of the type

$$\varphi(x_1, ..., x_m) \rightarrow R(y_1, ..., y_r),$$

where $\varphi(x_1, ..., x_m)$ is a conjunction of predicates on variables $x_1, ..., x_m$, possibly containing internal disjunctions, and $R(y_1, ..., y_r)$ denotes a target concept of arity r, where $y_1, ..., y_r$ is a subset of $x_1, ..., x_m$. In the HDL, the range-restriction bias is assumed, i.e. the variables occurring in the head of a clause must also occur in the body. The complete description of a concept usually consists of a disjunction of a set of inductive hypotheses.

Let us now consider the operational mechanism supporting the evaluation of a hypothesis in HDL. As previously mentioned, the learning instances are stored in database relations. Then, considering the inductive hypothesis $\varphi(x_1, ..., x_m) \rightarrow R(y_1, ..., y_r)$, to check for its consistency (completeness) means to verify that the extension of $\varphi(x_1, ..., x_m)$, projected on the variables $\{y_1, ..., y_r\}$, is included into (includes) the extension of $R(y_1, ..., y_r)$. The extension of $\varphi(x_1, ..., x_m)$ can be computed by applying a sequence of selection and natural join operators, as usual in relational databases.

Every conjunctive formula in the considered HDL can be evaluated with a similar procedure. Notice that, doing in this way, the object properties necessary to the selection operator are computed on demand, exiting so from the scheme of pure relational algebra.

3. A FRAMEWORK FOR LEARNING IN HDL

In this section we will show how the HDL language, defined in the previous section, lends itself to the use of induction algorithms with complementary abilities. First of all, we will briefly recall how inductive inference, specified via a high level strategy, can be realized on a database by means of *relational algebra*. The main advantage of this approach, already discussed in [8] and applied in various systems, is the possibility of directly embedding learning abilities into a database manager. This possibility, which is important for data mining applications, is a common platform for all the learning approaches exploiting the described HDL, and allows a natural cooperation among them. The considered approaches are the symbolic one, exploited in the systems SMART+ [9] and WHY [10,11], the genetic one, in REGAL [17, 18], and the connectionist one, in SNAP [23].

A multistrategy cooperative framework organizes learning. At the highest level, a symbolic relational learner like SMART+ is responsible for constructing a first order classification theory. The genetic and connectionist strategies work at the lower level, by refining the internal disjunctions occurring in the hypotheses constructed by the symbolic learner. In the following, we will describe how the above mentioned learning systems, even though requiring different internal representations, can effectively share the same HDL, and, in particular, how they can all deal with internal disjunction, whose advantages have been underlined in the previous sections. Moreover, we will describe methods for easily specifying, for each approach, declarative biases, both syntactic and semantic, when they are needed. Finally, we will show how the introduced representation framework offers the means of handling in a uniform way both discrete and continuous attribute values occurring in the description of the examples.

3.1 The Relational Learning Component

The basic learning strategy considered in this paper is a top-down construction of a hypothesis tree: the nodes of the tree are hypotheses (i.e., formulas that represent intensional descriptions of relations), whereas the edges are labeled by inductive operators. The growth of the tree along a branch stops when either a consistent hypothesis is found or when the formula associated to the leaf satisfies some heuristic halt criterion. Every time a new formula is created and added to the tree, the corresponding extensional relation is created as well, and it is stored on a mass storage device. In this way the search algorithm keeps checkpoints of the states it went through, thus allowing more sophisticated search strategies, requiring a degree of backtracking, to be used. As a side-effect, the induction algorithm is robust, as it can easily run for long periods, if necessary, surviving possible operating system crashes.

The search tree is generated step by step, and the following actions are repeated at each step:

• Select from the frontier of the tree a set Φ of formulas φ that are good candidates for specialization by addition of new literals.

- Determine the set of literals P which can be added to each φ in Φ according to domain knowledge and heuristic criteria.
- Generate a set Ψ of new formulas ψ by AND-ing each φ in Φ with a literal p in P, selected according to some heuristic evaluation.
- For every ψ in Ψ that is consistent, declare covered the instances verifying ψ, update the frontier and restart.

The process stops when either all positive instances are covered or no more promising formulas are in the frontier. Detailed examples of hypothesis tree generations can be found in [6]. As it appears from the algorithm, many heuristics and biases are involved, which have been widely investigated in other papers. In principle, any heuristics suggested in the literature can be used within this framework without any special restriction. Moreover, since the software structure is very robust and capable of dealing with large datasets, also other search strategies, such as best-first or hill-climbing, can be implemented, instead of the beam search strategy discussed in the algorithm.

The use of relational algebra parallels, at the extensional level, the intensional search algorithm, and generates a tree, whose nodes contain the extensions of the relations associated to the corresponding intensional nodes. This approach requires large storage resources, whereas it may reduce the time to evaluate the generated hypotheses. On the contrary, by avoiding keeping the extensional relations, no large space resources are needed, but the time to evaluate new hypotheses, made on demand, may be much larger.

Another aspect characterizing the present framework that needs to be explained is the way in which internal disjunction is dealt with. In particular, we have to face the problem of evaluating a literal as a candidate to be added to a formula φ. This can be seen as a specific learning subtask. In the following subsection, solutions to this subtask will be described. All of them use the notion of *predicate template*, which plays a role similar to that of *schemata* [24].

Given a predicate p, containing an internal disjunction, the *template* T_p associated to p, is the maximum internal disjunction allowed for p: any internal disjunction occurring in an instance of p can only be a subset of T_p. As an example, the template for the predicate "Color" shall enumerate all the considered color values. If the attribute has a continuous range of values, as, for instance, the predicate "Height", its template shall denote the maximum range allowed for the values.

3.2 Learning Internal Disjunctions Locally

In the general-to-specific learning strategy described above, it has been assumed that an algorithm for determining the best predicate to add to a formula φ is available. When a candidate predicate contains an internal disjunction as one of its arguments, the "best" disjunction among all the possible ones is to be determined. For instance, in the predicate Color(x, [yellow, red, blue]) one of the possibilities is Color(x, [yellow, blue]). We designed algorithms to learn both discrete disjunctions and continuous intervals. These algorithms determine internal disjunctions by performing a local optimization, which accounts only for the literals already present in the current hypothesis. On the other hand, a global optimization of all internal disjunctions occurring in the predicates of a formula requires excessive computational resources, if dealt with classical optimization algorithms. However, the problem can be approached by using a Genetic Algorithm (GA) as a search engine [9].

In [17] we showed how a formula with predicates containing an internal disjunction can be mapped to a bitstring, processable by a GA. The way internal disjunctions are encoded in a chromosome needs some explanation. In the case of a language template containing only discrete sets, it is immediate to map the elements of the different sets to a bit string. In fact, every item in a template corresponds to a boolean gene on the chromosome. When the gene assumes the value 1, the corresponding attribute value will be considered present in the corresponding internal disjunction, otherwise not. When an individual needs to be evaluated, the actual values for the internal disjunctions are loaded from the bit string into the template so creating a new formula which is matched on the learning instances.

For a continuous interval, defined by a pair of reals, a different approach has been adopted. A first solution, in agreement with the most classical GA approach, is to transform the real numbers into discrete integers that can be encoded as binary numbers; in this way, the chromosome becomes a bit string. This kind of transformation has been used in a previous version of SMART+ [9]. A different approach can be taken, following the tendency of the Evolutionary Computation approach, using a chromosome composed of "real-valued" genes.

Global optimization is used in two ways. Discrete optimization of internal disjunctions is actually the learning strategy of REGAL: learning is nothing else that a search for the "best" internal disjunctions in a language template. Continuous optimization of intervals is performed by SMART+ as a postprocessing of the formulas learned by the symbolic learner and containing numerical parameters. A few comments are in order regarding the template. The fixed length of the template is not a limitation on the type of hypotheses can be learned. In fact, the formula φ, encoded in the template, may be generated by a symbolic learner, such as SMART+, which uses the full language defined in Section 2. As a special case, the template can be supplied directly by the user, when he/she has sufficient knowledge to establish the maximum reasonable complexity of the sought descriptions.

3.3 Dealing with Uncertainty and Refining a Theory by Performing the Quadratic Error Gradient Descent

A frequent problem in real world applications is the presence of uncertainty in the data. Boolean features, as they are learned by symbolic systems, have little reliability because of the intrinsic variability of the phenomena they try to capture. In this sense, the neural network approach is considered much more effective, even if the networks look like black boxes, difficult to understand to the end user. An attempt to combine the properties of neural networks with the ones of symbolic systems is represented by continuous valued logics, such as Fuzzy Logic and Probabilistic Logic, which have been successful in some applications. Nevertheless, the real power of neural networks, not yet available for continuous valued logics, derives from effective learning algorithms, such as back-propagation [26]. For this reason several methods have been proposed, trying to extend learning algorithms in such a way that the quadratic error gradient descent can be applied, as it has been done for propositional logic with neural networks [27, 23].

Here we will follow a similar approach, but starting from the First Order Logic language we have used so far, instead of propositional logic. The aim is to exploit the abstraction properties offered by this framework in order to learn better and more compact descriptions [28].

4. DISCUSSION

In the previous sections we have presented a framework for integrating multiple learning strategies, which is characterized by two basic choices:

(a) a separation between the (logical) hypothesis description language, and the (object oriented) instance description language,
(b) the adoption of the annotated predicate calculus, as hypothesis description language, with internal disjunction as basic construct.

It is worth noting that the two choices are independent, even though they nicely combine together. On the other hand, the first choice is, to a large extent, motivated by the need of accessing data in large databases. The representation framework, as it has been described in this paper, is not comprehensive of all the facilities that it can supply; in fact, the description only presents those features that can be used at the same time by a multistrategy approach including the symbolic, genetic and connectionist paradigms. More specifically, the biases of the core framework, and the possible extensions, can be summarized in the following. A brief comparison with alternative approaches is also sketched.

Instances are represented as tuples (objects with attributes) in a database. Properties can be defined on the objects either by means of semantic functions or directly by providing the corresponding relations in extensional form.

Moreover, a set of operational predicates [19] is supplied. These predicates can be constraint predicates, with no internal disjunction, or learnable predicates, with one term in disjunctive form. The semantic of the predicates is defined in terms of properties, and negation of atoms is allowed. A well-formed HDL formula is any range-restricted, non recursive, normal clause. Starting from this core, both stronger and weaker biases can be used, depending on the specific learning task. For instance, in the system SMART+ existential and numerical quantification [20] is allowed. On the other hand, a form of determinacy (extensionally controlled) can be imposed for limiting the growth of too large relations.

In the core framework presented here there is no background knowledge, at least formally. Actually, object properties, represented as semantic mappings, are a kind of background knowledge; they could be represented as Horn clauses and used deductively. On the other hand, complex mechanisms to handle possibly large bodies of domain theories have been implemented and used, integrating induction and deduction, both in [14] and in SMART+. Moreover, in the system WHY, also a causal model of the domain can be used in an abuctive way [10,11].

In the basic framework, several search methods and strategies can be applied. In particular, inductive hill-climbing, best-first, depth-first and beam search, controlled by a variety of heuristic evaluations [9], can be used in SMART+. Genetic search in used in REGAL, and backpropagation or Delta-rule are used in SNAP [23]. Keeping separate by a functional abstraction layer the extension (database) from the intension (learned knowledge) offers the advantage that efficient mechanisms for extensional relations can be designed by exploiting the database technology. The method becomes particularly effective when object properties have a closed-form definition: in this case most of the intermediate relations, necessary to compute the extension of a hypothesis, do not need to be created; they can be obtained, instead, on demand by evaluating the properties of objects bound to the variables.

Another benefit is the flexibility at the application level. When the user wants to introduce or delete a feature, it is not necessary to do any preprocessing on the dataset, but it is sufficient to define/delete the properties necessary to compute the new feature. New properties can also be defined on-line by the learning system itself, if a constructive learning strategy is used. In this way, more robust learners can be designed. On the other hand, this approach has the drawback that the constants defined in the dataset cannot directly occur in the logical formulas, but they have to be manipulated through suitable objects. Moreover, single instances cannot be accessed directly, but only through the general operators which are applicable to the relations. However, even taking advantage of such solutions, it is true that the presented framework may be quite inefficient when one must work not on whole relations but on single items, such as constants or instances.

REFERENCES

1. Plotkin, G. (1971). "A Further Note on Inductive Generalization". *Machine Intelligence 6*, 101-124.
2. Hayes-Roth, F. and McDermott , J. (1978). "An Interference Matching Technique for Inducing Abstractions". *Communications of the ACM, 21*, 401-411.
3. Michalski R.S., Chilauski R.L. (1980): "Learning by being told and learning from examples: an experimental comparison of the two methods of knowledge acquisition in the context of developing an expert system for soybean disease diagnosis". *International Journal of Policy Analysis and Information Systems, 4*, 125-126.
4. Kodratoff Y., Ganascia J.G. (1986): "Improving the generalization step in learning". In J.G.Carbonell, R.S. Michalski and T. M. Mitchell (Eds.), *Machine Learning: An Artificial Intelligence Approach, Vol. II*, Morgan Kaufmann (San Mateo, CA), pp. 215-244.
5. Bergadano, F., Giordana, A., Saitta L. (1988): "Automated concept acquisition in noisy environments". *IEEE Transactions on Pattern Analysis and Machine Intelligence, PAMI-10* , 555-577.
6. Bergadano, F., Giordana, A., Saitta L. (1991). *Machine Learning: An Integrated Approach and its Application,* Ellis Horwood, Chichester, UK.
7. Pazzani M. and Kibler D. (1992). "The Utility of Knowledge in Inductive Learning". *Machine Learning, 9,* 57-94.
8. Giordana A., Saitta L., Bergadano F., Brancadori F., De Marchi D. (1993). "ENIGMA: A System that Learns Diagnostic Knowledge". *IEEE Transactions on Knowledge and Data Engineering, KDE-5,* 15-28.
9. Botta M. and Giordana A. (1993). "Smart+: a Multi Strategy Learning Tool". *Proc. 13th Int. Joint Conference on Artificial Intelligence* (Chambéry, France), pp. 937-943.
10. Baroglio C., Botta M. and Saitta L. (1994). "WHY: A System that Learns Using Causal Models and Examples". In R. Michalski & G. Tecuci (Eds.), *Machine Learning: A Multistrategy Approach, Vol. IV,* Morgan Kaufmann, San Francisco, CA, pp. 319-347.
11. Saitta L., Botta M., Neri F. (1993): "Multistrategy Learning and Theory Revision", *Machine Learning, 11,* 153-172.

12. Quinlan, J. R. (1990). Learning Logical Definitions from Relations. *Machine Learning, 5,* 239-266.
13. Muggleton S. (1991). "Inductive Logic Programming". *New Generation Computing, 8,* 295-318.
14. Bergadano, F. & Giordana, A. (1990). Guiding Induction with Domain Theories. In R. S. Michalski and Y. Kodratoff (Eds) *Machine Learning: An Artificial Intelligence Approach, Vol. III,* Morgan Kaufmann (Los Altos, CA), pp. 474-492.
15. Haussler, D. (1989). "Learning Conjunctive Concepts in Structural Domains". *Machine Learning, 4,* 7-40.
16. Ade H., De Raedt L. and Bruynooghe M. (1995). "Declarative Bias for Specific-to-General ILP Systems". *Machine Learning,* Special Issue on Declarative Bias.
17. Giordana A. & Neri F. (1996). "Search Intensive Concept Induction". *Evolutionary Computation, 3 ,* 375-416.
18. Neri F., Saitta L. (1996): "A Study of the *Universal Suffrage* Selection Operator". *Evolutionary Computation, 4,* 87-107.
19. Mitchell T. , Keller R. and Kedar-Cabelli S. (1986). "Explanation-Based Generalization". *Machine Learning, 1,* 47-80.
20. Michalski R.S. (1980). "Pattern recognition as a rule-guided inductive inference". *IEEE Trans. on Pattern Analysis and Machine Intelligence, PAMI-2,* 349-361.
21. Michalski R. S., Carbonell J. G., Mitchell T. (Eds.) (1983): *Machine Learning: An Artificial Intelligence Approach, Vol. I,* Tioga Publishing Co. (Palo Alto, CA).
22. Giordana A., Neri F., Saitta L. and Botta M. (1997). "Integrated Learning in First Order Logic: A General Framework". *Machine Learning.* To appear.
23. Botta M. and Giordana A. (1996). "Combining Symbolic and Numeric Methods for Learning to Predict Temporal Series". *Proc. 3rd Int. Multistrategy Learning Workshop* (Harpers Ferry, WV), pp. 125-134.
24. Morik K., Wrobel S., Kietz J. U., Emde W. (1993). *Knowledge Acquisition and Machine Learning - Theory, Methods, and Applications,* Academic Press.
26. D.E. Rumelhart, J.L. McClelland and the PDP Research Group (Eds.), *Parallel Distributed Processing: Exploration in the Microstructure of Cognition,* MIT Press, Cambridge, MA
27. Towell G.G. and Shavlik J.W. (1994). Knowledge-Based Artificial Neural Networks. *Artificial Intelligence, 70,* 119-165.
28. Blanzieri E. and Katenkamp P. (1996). "Learning Radial Basis Function Networks On-Line". *Proc. 13th Int. Conf. on Machine Learning* (Bari, Italy), pp. 37-45.

MACHINE LEARNING AND CASE-BASED REASONING: THEIR POTENTIAL ROLE IN PREVENTING THE OUTBREAK OF WARS OR IN ENDING THEM

R. Trappl

Austrian Research Institute for Artificial Intelligence, Vienna, Austria
and
University of Vienna, Vienna, Austria

J. Fürnkranz and J. Petrak

Austrian Research Institute for Artificial Intelligence, Vienna, Austria

J. Bercovitch

University of Canterbury, Christchurch, New Zeland

Abstract. In a current project we investigate the potential contribution of Artificial Intelligence for the avoidance and termination of crises and wars. This paper reports some results obtained by analyzing international conflict databases using machine learning and case-based reasoning techniques.

1 Introduction

Research in Artificial Intelligence has always been heavily supported by "defense agencies". While enormous amounts of money have been and still are spent on the development of AI methods for military purposes, practically no effort is undertaken to use these methods to support the *prevention and termination* of conflicts and wars. We believe that Artificial Intelligence has a great potential for peacefare although research in this area has not yet received the attention that it deserves (Trappl, 1986, 1992).

One possible contribution of AI to peacefare is the knowledge that can be gained by analyzing databases of international conflicts or conflict management attempts with machine learning algorithms. This paper presents two case studies for such an approach.

2 International Conflict Databases

In recent years the importance of empirical studies has also been recognized in the *international relations* research community. This increasing interest in empirical methods was a major source of motivation for the development of a variety of databases that try to capture the important aspects of international *crises and code them into suitable attributes*. The most important among these databases are:

---
* Corresponding author. E-mail: `robert@ai.univie.ac.at`

- the Correlates of War Militarized Interstate Disputes dataset (Gochman & Maoz, 1984),
- the International Crisis Behavior (ICB) project (Brecher, Wilkenfeld, & Moser, 1988; Wilkenfeld, Brecher, & Moser, 1988),
- the COPDAB dataset (Azar, 1980),
- the event data sets of the KEDS and PANDA projects (Schrodt & Davis, 1994; Vogele, 1994; Bond, Bennet, & Voegele, 1994),
- the Butterworth dataset (Butterworth, 1976),
- the SHERFACS database (Sherman, 1988),
- the KOSIMO database of conflicts (Pfetsch & Billing, 1994),
- the CONFMAN database of conflict management attempts (Bercovitch & Langley, 1993).

We believe that databases like the above-mentioned provide a promising application area for symbolic machine learning algorithms. This study reports the results of such an attempt. After considering the availability and scope of these datasets, we eventually chose to work with the CONFMAN and KOSIMO databases.

2.1   The CONFMAN Database

The development of the *International Conflict Management (CONFMAN) Dataset* is a project that is conducted under the supervision of Jacob Bercovitch at the Department of Political Science of the University of Canterbury, New Zealand. This database is of particular interest for Machine Learning research, as it has been generated with the explicit aim of empirical analysis. Its primary focus is international mediation. Its aim is to both further our understanding of mediation, and facilitate the comparative investigation of different conflict management mechanisms.

Prompted by dissatisfaction with previous studies, which have rested on ideo-graphic or normative approaches, the development of this database was begun with the aim of furthering the much needed empirical investigation of conflict management within a sound theoretical framework. The project is founded on the contingency approach to the study of international conflict management which regards the outcome of management efforts as contingent upon a number of contextual and process variables. The contingency approach encourages systematic empirical research because it recognizes variables and attributes with explicit operational criteria.

The database should facilitate the answer to such fundamental questions as "How do international mediation, and other forms of conflict management work?" and "Under what conditions are respective conflict management efforts most effective?". In answering these questions it is hoped the project will make a concrete contribution to the improvement of the international conflict management process.

A mediation attempt is defined as the formal or institutionalized non-violent and non-judicial intervention of an outsider or third party willing to help both

Table 1. Attributes of the CONFMAN database

Attribute	Type	Description
V1	numeric	Dispute Number
V2/V3	ordinal/numeric	Duration (grouped/raw)
V4/V5	ordinal/numeric	Fatalities (grouped/raw)
V6	ordinal	Dispute Intensity
V7	ordinal	System Period
V8	nominal	Geographic Region
V9–V11	nominal	Issue 1 – Issue 3
V12	nominal	Final Outcome
V13	nominal	Dispute Initiator
V14/V15	nominal	Identity Party A/B
V16/V17	nominal	Time in IS A/B
V18	nominal	Alignment
V19/V20;V21/V22	numeric;ordinal	Power A/B (raw;grouped)
V23	nominal	Previous Relation
V24/V25	nominal	Political System A/B
V26/V27	numeric	Number of Parties A/B
V28/V29	ordinal	Homogeneity A/B
V30/V31	ordinal	Political rights A/B
V32/V33	ordinal	Civil liberties A/B
V35	nominal	Conflict Management Type
V36	nominal	Third Party Identity
V37	nominal	Mediator Rank
V38	nominal	Mediation Strategies
V39	nominal	Previous Relationship
V40	ordinal	Prev Attempts
V41	ordinal	Prev Attempts this Mediation
V42/V43	ordinal/numeric	Timing (grouped/raw)
V44	nominal	Initiated by
V45	nominal	Environment
V80	ordinal	m intensity
V70	ordinal	power disp
V75/V76	nominal	Human Rights A/B
V77	ordinal	Human Rights Disparity
V91	nominal	Political System Type
V81	binary	Political System Difference
V92	ordinal	Ally Numbers
V82	binary	Ally Support Disparity
V93	nominal	Homog. Type
V83	nominal	Homog. Comparison
V94	ordinal	Time in System
V84	binary	Time in System Comparison
V90	nominal	Total Issues
V46	nominal	Outcome (detailed)
V99	binary	Outcome (binary)

disputants seek an acceptable outcome. An offer of mediation services is included in this understanding of an intervention. Other forms of conflict management that are encompassed are negotiation, arbitration/adjudication, multilateral conference, and referral to an international organization. The referral of a dispute to an international organization is coded as a separate event from any subsequent mediation or adjudication by that organization.

The central task of this research project has been the compilation of an extensive original dataset of international conflict management events since 1945.

Primary information sources included Keesings Contemporary Archives (laterly Keesings Record of World Events), The Times Index, and The New York Times Index. Whenever necessary more detailed contemporary press reports or reputable historical accounts were also utilized.

The dataset that was used in the current study encompasses 921 international disputes and management attempts from 241 disputes since 1945. The attributes we used are listed in Table 1. This database — or previous versions of it — has been analyzed extensively with statistical methods, most recently in (Bercovitch & Wells, 1993; Bercovitch & Houston, 1993; Bercovitch & Lamare, 1993; Bercovitch & Langley, 1993).

2.2   The KOSIMO Database

The KOSIMO database has been developed under the supervision of Frank Pfetsch at the Institute of Political Science at the University of Heidelberg, Germany (Pfetsch & Billing, 1994). The database is an attempt to unify and extend case lists and databases of several previous research projects: primarily (Butterworth, 1976),(Brecher et al., 1988),(Wilkenfeld et al., 1988), but also (Gantzel & Meyer-Stamer, 1986), (Holsti, 1983), and others. It contains more than 1400 conflicts from 1482 to 1990 encoded in three tables. We have concentrated on the NOPUTSCH table that describes 547 internal and international conflicts and wars between 1945 and 1990. As the database was not generated with the explicit aim of empirical analysis, its complex structure (list-valued, hierarchical, textual, and multi-dimensional fields) made several transformation necessary, which are described in more detail in (Fürnkranz, Petrak, & Trappl, 1997). As many of the above-mentioned problems can not trivially be solved by attribute-value based machine learning algorithms, we tried to analyze the database with case-based techniques.

3   Inductive Learning Techniques

As the simple attribute-value format of the CONFMAN database suggests the use of decision tree learning algorithms, we first tried to analyze it with C4.5 (Quinlan, 1993). We started with investigating the predictive accuracies that can be obtained by learning trees pruned to different degrees. Figure 1 shows the results in terms of accuracy on the training set (purity) and predictive accuracy as estimated by a 10-fold cross-validation for various choices of C4.5's -m and -c parameters.

The performance of C4.5 peaks at parameter settings -m 30 and -c 10, i.e. when C4.5 ensures that only those decision nodes are further expanded for which at least two children cover more than 30 training examples or when a relatively high degree of pruning is employed.[4] The default accuracy of this domain (indicated by the flat lines at the bottom of Fig. 1) is 56.8%, the result for C4.5's

---

[4] The value range of the -c parameter in C4.5 is from 0 to 100, where a low value indicates heavy pruning, while a value of 100 indicates no pruning.

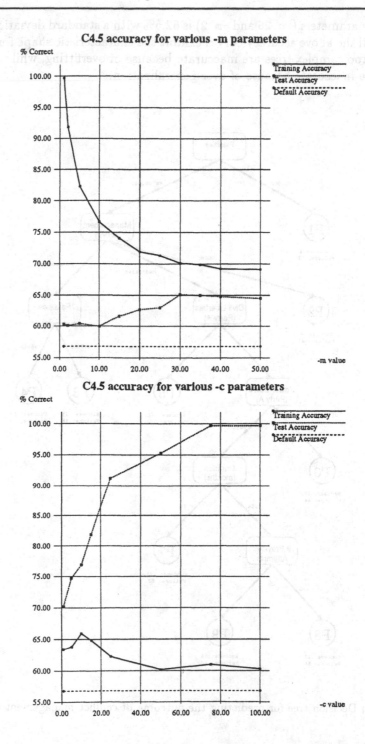

Fig. 1. Results for various settings of C4.5's -m and -c parameters

default parameters (-c 25 and -m 2) is 62.5% with a standard deviation of ±5.2. All in all the above accuracy curves exhibit the characteristic shape for noisy domains: too complex trees are inaccurate because of overfitting, while too simple trees are inaccurate because of over-generalization.

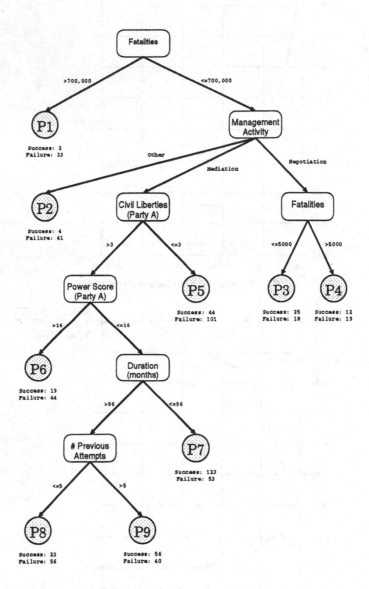

Fig. 2. Decision tree for predicting the outcome of conflict management attempts

We have also tested the combination of the best two values found above, which resulted in a decision tree with an estimated predictive accuracy of around

66.7% (±3.7). The tree obtained by this parameter setting is shown in Fig. 2. Three rules (P3, P7, and P9) cover a majority of successful conflict management attempts. In particular rule P7 which describes 176 attempts, among them 123 (69.9%) successes, looks interesting.

However, a closer investigation of trees like the one of Fig. 2 shows that the quality of the rules in a tree is very unstable. Some rules discriminate very well between successes and failures, while other leaves seem to have about the same distribution of successes and failures as in the original datasets and can thus not be expected to perform better than mode prediction. Consequently we switched to examining single rules instead of entire trees. In a first experiment we generated a decision tree using C4.5 and examined it for nodes containing only successful or only unsuccessful mediation outcomes. A typical example of such a pure rule is shown below (a complete listing of the rules that have been discovered by this process is given in appendix A).

> If there have been less than 400 fatalities and
>   party B's raw power index is not extremely high and
>   the conflict management type was mediation and
>   the conflict lasted between 1 and 3 months
> then the conflict management was always successful
>   in 15 mediation attempts in 8 different conflicts.

In a special experiment we tried to answer the question, what factors influence the success or failure of a given mediation strategy. For this purpose we generated decision trees in which we forced C4.5 to use the attribute *Mediation Strategy* at the root, so that it will try to find the best sub-tree discriminating between successful and unsuccessful conflict management attempts for a given mediation strategy. As the quality of the leaves of the found trees varied, we again examined the trees for leaves in which successes or failures significantly dominate the matching mediation attempts. One such rule was

> If the mediator has mixed relationships with the conflict parties,
>   or is from the same block as both of them,
>   or from a different block as both of them,
>   and
>   the mediation environment is party B's territory,
>   a third party's territory or a composite
> then directive mediation strategies have been
>   successful in 37 mediation attempts and have
>   failed in 7 mediation attempts.

A complete decision tree resulting from these experiments can be found in (Fürnkranz et al., 1997).

In another experiment we tried to compare the results of *feature subset selection* with the identification of relevant factors by statistical analysis. For this purpose we used the feature subset selection procedure implemented in an early version of the publicly available machine learning library MLC++ (Kohavi, John, Manley, & Pfleger, 1994), which realizes a wrapper approach around

C4.5 (John, Kohavi, & Pfleger, 1994). It starts with an empty set of attributes and greedily adds the attribute that gives the highest increase in estimated predictive accuracy for the tree grown from the new set of attributes. Alternatively, the algorithm can also choose to delete an existing attribute from the current set of attributes. Predictive accuracy is estimated with consecutive 10-fold cross-validation experiments (with different random splits) until the standard deviation of the resulting estimate is below 1%. If no feature can be added or deleted without decreasing the estimated accuracy of the tree for two consecutive tries, the program stops with the current set of features. In order to avoid to be too short-sighted a one-time decrease is not sufficient for stopping the algorithm. In this case two features may be added at a time if this increases accuracy.

Table 2. Relevant attributes detected by feature subset selection

Choice	Variable	Purity	# X-vals	Accuracy
	C4.5 -m 2 -c 25			
1	V39 Previous Relationship	64.2%	3	63.2%
2	V20 Raw Power Score B	68.1%	4	65.5%
3	V37 Mediator Rank	71.7%	4	65.9%
4	V27 Number of Parties B	74.1%	4	67.0%
5	V22 Grouped Power Score B	73.4%	4	67.3%
6	V06 Dispute Intensity	—	—	—
6	V45 Environment	70.3%	4	67.5%
8	V11 Issue 3	70.5%	4	67.7%
	C4.5 -m 2 -c 25 -s			
1	V39 Previous Relationship	64.3%	2	63.3%
2	V19 Raw Power Score A	67.5%	2	65.4%
3	V27 Number of Parties B	70.3%	4	65.8%
4	V10 Issue 2	70.6%	4	66.7%
5	V70 Power Disparity	—	—	—
5	(V82 Ally Support Disparity)	71.3%	3	66.9%
7	V90 Total Issues	71.3%	3	67.1%
8	V21 Grouped Power Score A	—	—	—
8	V45 Environment	70.9%	4	67.1%
10	(V82 is deleted again)	70.9%	3	67.7%

Table 2 summarizes the output of MLC++ from two experiments, one using the default settings for the parameters, and one with the -s option turned on, which allows C4.5 to lump different outcomes of a test together thereby obtaining simpler trees. We have tried a few different parameter settings, in particular those that yielded the best results in previous experiments (Fig. 1). However in this case, the default choices seemed to be very good, which indicates that only relevant attributes are used and that therefore too high settings of the -m parameter and too low settings of the -c parameter may force C4.5 to throw away relevant information. For each of the two experiments we report the purity of the final tree, the number of cross-validations needed to get the standard deviation below 1% and most importantly the estimated accuracy of the tree. The tables have to be read from the top to the bottom.

The final decision tree in both cases consisted of 8 variables and had an accuracy of above 67%. It is interesting that in the experiment where the -s parameter was activated, the program at one point had 9 variables in the tree, but the feature concerning the disparity of the support of each party's allies could be deleted again at the end with a further increase of accuracy. This shows that the algorithm does not necessarily converge towards an optimal subset of features. It may for example be the case that adding a combination of certain attributes yields a better tree, while adding only one of them results in a worse tree. The analysis reveals that for example the attributes chosen in the tree of Fig. 2 are very different from the attributes that appear in Table 2 which have produced a more accurate tree.

It is interesting to compare the above results with the results produced with classical statistical methods. Table 3 shows the most relevant aspects of mediation attempts that have been identified in previous work (Bercovitch & Lamare, 1993). There is obviously a considerable overlap. Almost all of the variables of Table 3 appear in one of the two experiments of Table 2, most of them in both. The most notable exception is the absence of mediation strategy. The number of fatalities, which is also not considered by Machine Learning, is partially reflected in the intensity of the conflict, which has been recognized as important, although only in one experiment. Using only fatalities for generating a decision tree would only yield 62.4% accuracy using the same grouping as in (Bercovitch & Lamare, 1993). On the other hand, the Machine Learning method has attributed a higher significance to the previous relation of the mediator (63.3%). In addition aspects concerning the power of the conflict parties and about the number of parties involved on each side have been considered.

Table 3. Relevant features for mediation outcome detected by statistical analysis

Fatalities
Mediation Environment
Mediation Strategy
Previous Relations of Mediator
Issues
Mediator Rank

4   Case-Based Techniques

Situations of international conflict and war, like other complex human life situations, are often described and explained in terms of previous similar situations. Such comparisons often help to understand the various possibilities of actions the participants and international organizations can choose, and their possible consequences. Similarity-based case retrieval and analysis can therefore be a useful tool for analyzing a new conflict situation. This and the fact that the KOSIMO database was less susceptible to analysis with machine learning techniques, because it was not coded in a strict attribute-value fashion, but included list-valued, multi-dimensional and hierarchical fields motivated our experiments with case-based learning and similarity-based case retrieval techniques. These experiments were performed with the VIE-CBR tool box, a flexible and extensible library of Common LISP routines that allows easy experimentation with alternative algorithms for any of its functional components (Petrak, 1994).

Sim	Year	Conflict
		Bosnia-Herzegovina
0.62	1938	Germany-Czechoslovakia (Munich Treaty)
0.60	1948	Israel I (Palestine War)
0.57	1974	Cyprus IV (Turkish Invasion)
0.55	1965	India XVI (Kashmir IV)
0.54	1968	CSSR (Invasion)
		Germany-Czechoslovakia (Munich Treaty)
0.77	1968	CSSR (Invasion)
0.75	1953	GDR (17. June 1953)
0.72	1946	Greece (Civil War II)
0.67	1948	Berlin I (Blockade)
0.66	1961	Berlin III (Wall Erection)
		USA-Grenada
0.66	1959	Dominican Republic I (Intervention)
0.57	1962	Cuba IV ('Cuba-Crisis')
0.57	1954	Guatemala I (Intervention)
0.57	1973	Libya-USA
0.57	1945	Triest

Fig. 3. The five best matches for three selected cases ordered by decreasing similarities (English translation of the original German KOSIMO database entries)

First, we simply tried to retrieve the most similar cases to a given conflict from the KOSIMO database. Figure 3 shows the retrieval of the five nearest neighbors of three selected cases in the database when using a similarity measure previously defined by a domain expert. The case "548 Bosnia-Herzegovina" has been coded and added to the database by one of the authors of the KOSIMO

Table 4. Error rates and output similarities of 1-NN for KOSIMO

1-NN	ERGEBNISM		ERGEBNIST		ERGEBNISP		LOESUNG		INTENS	
SIM-EVEN	49%	0.62	44%	0.62	80%	0.41	70%	0.40	54%	0.78
SIM-EXPERT	47%	0.65	45%	0.61	82%	0.39	73%	0.38	55%	0.78
Default	49%	0.51	37%	0.63	98%	0.38	73%	0.27	65%	0.70

database for this experiment. It seems remarkable that one of the two situations (Vietnam and the Munich Treaty) with which the conflict in Bosnia was often compared by politicians before deciding to intervene or not, was ranked the most similar case by the program. This and several other similar experiments have been discussed at the *Second International Workshop on the Potential Contribution of Artificial Intelligence to the Avoidance of Crises and Wars* in Vienna and the invited domain experts (among them the creators of the database) found meaningful explanations for the similarity of the retrieved cases.

We have also performed experiments that aimed at deriving automatic classification of unknown cases using nearest neighbor techniques. Table 4 shows the results obtained by trying to predict the military (ERGEBNISM), territorial (ERGEBNIST), political (ERGEBNISP) results of a conflict, the type of conflict resolution used in its settlement (LOESUNG), and the intensity of the conflict (INTENS). We have tried two different similarity measures, one that gives equal weights to all features (SIM-EVEN), and one that was provided by one of the creators of the KOSIMO database (SIM-EXPERT). The results in the first column show the error rates, while the numbers in the second column show a *output similarity* measure that tries to take into account the distance between the predicted output value and its actual value in the case. All numbers shown are the average of 10 cross-validated runs.

Most of the error rates are only slightly better than default accuracy. The best improvements in error rate were achieved for ERGEBNISP and INTENS, while the algorithm performed badly for predicting territorial outcomes of a conflict (ERGEBNIST). Note, however, that for these experiments, missing values were treated like other values. All cases that had a missing classification but were classified with a non-missing value, received output similarity zero. For the field ERGEBNIST, for instance, no value was specified in 63% of all cases in KOSIMO. This might be an explanation why the results for this field were particularly bad.

The results are a little better when we look at the output similarities of the 1-NN predictions. Except for the ERGEBNIST experiments they are always substantially better than mode prediction, which suggests that the retrieved cases capture some aspects that are relevant for the classification of the current case. This is illustrated in Fig. 4 which shows with which relative frequency the output similarities for field INTENS occurred when applying 1-NN with SIM-EVEN to KOSIMO (first line of Table 4). It is obvious that 1-NN's predictions are closer to the target than mode prediction. In about 90% of the cases 1-NN

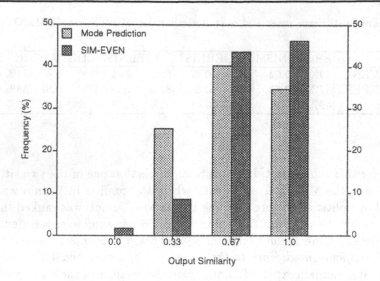

Fig. 4. Output similarity of field INTENS (mode prediction vs. SIM-EVEN)[5]

correctly predicts the intensity level or misses only by one degree.

We have also tried several enhancements of the basic nearest neighbor algorithm, like e.g. case weighting or considering more than one neighbor for the derivation of a classification. These and more experiments with different weighting schemes and on a different subset of the KOSIMO database can be found in (Fürnkranz et al., 1997) and (Petrak, Trappl, & Fürnkranz, 1994).

5   Related Work and Discussion

There has been some previous work on rule induction via decision tree learning, which we will briefly discuss below. More extensive overviews can be found in (Mallery, 1988), (Schrodt, 1991a), and (Schrodt, 1997).

Schrodt (1991b) has performed several experiments in predicting interstate conflict outcomes using the Butterworth "Interstate Security Conflicts, 1945–1974" (Butterworth, 1976). He used his own implementation of ID3, the predecessor of C4.5, to learn decision trees for predicting the effects of management efforts with respect to five different outcomes. In all his experiments the estimated predictive accuracy of the learned trees was below mode prediction accuracy, i.e. below the accuracy that one would achieve by always predicting

---

[5] The field INTENS describes the intensity of the conflict on a value range from 1 to 4. The output similarity measure can thus yield 4 different degrees of similarity from 0.0 (when 1 is predicted and 4 is correct and vice versa) to 1.0 (exact prediction) depending on the difference between the prediction and the actual intensity.

the majority class. However, his implementation of ID3 was not capable of dealing with numeric data and, more importantly, did not have C4.5's extensive pruning facilities. The only method used for getting simpler trees was manual feature subset selection by observing which attributes his algorithm typically selects near the root of the tree, which did not result in higher accuracies. Note that the method we chose for feature subset selection produces different results: for example the attributes chosen in the tree of Fig. 2 differ substantially from the attributes that appear in Table 2, which have been chosen to maximize predictive accuracy of the trees generated by C4.5, In our study simple decision trees are usually more accurate than an unpruned decision tree. However, even the unpruned tree exhibits a significant gain in predictive accuracy compared to mode prediction. Predicting the outcome of conflict management attempts thus seems to be an easier task than the prediction of aspects of the outcome of the conflict itself. A reason for this might be that conflict management events are more repetitive than the conflicts themselves.

Our results with C4.5 indicate that the quality of the rules inside the decision tree can vary substantially. In general, single rules can be more reliable than entire decision trees. Mallery and Sherman (1993) report a variety of rules that have been learned with I^2D (Unseld & Mallery, 1993), an improved version of ID3 that was specifically developed to deal with the structured nature of the SHERFACS dataset (Sherman, 1988). They report purity and coverage for the learned rules, but give no indication of their predictive accuracies. Estimating the predictive accuracy of single rules of a decision tree is more problematic than estimating the accuracy of the entire tree, because cross-validation can only estimate the accuracy of complete classifiers. The alternative, to reserve a subset of the data entirely for testing, is also not a good solution, because the size of these databases is in general fairly low compared to the number of attributes, so that an additional loss of training data would be detrimental to the quality of the learned rules. Schrodt (1991b) has used an entropy ratio, similar to the information score proposed in (Kononenko & Bratko, 1991) which gives a higher weight to correct predictions of rare classes.

6   Further Work

Currently we are working on a much larger and more recent version of the CONF-MAN database. Initial experiments with the C4.5 algorithm have promised a significant improvement of the results in terms of accuracy. We also plan to employ a wider range of machine learning and knowledge discovery techniques, e.g. the use of *inductive logic programming* techniques and the discovery of *partial determinations*. Another goal is to further improve the results by using domain-specific background knowledge. We also plan to address different sets of questions in the near future. An example for such a question would be "Has the set of factors that determine the success of a mediation outcome changed since the end of the *cold war*?". With recent versions of the CONFMAN database we hope to be able to shed some light on this question.

References

Azar, E. E. (1980). The conflict and peace data bank (COPDAB) project. *Journal of Conflict Resolution, 24*(1), 143–152.

Bercovitch, J., & Houston, A. (1993). Influence of mediation characteristics and behavior on the success of mediation in international relations. *The International Journal of Conflict Management, 4*(4), 297–321.

Bercovitch, J., & Lamare, J. W. (1993). The process of international mediation: An analysis of the determinants of successful and unsuccessful outcomes. *Australian Journal of Political Science, 28*, 290–305.

Bercovitch, J., & Langley, J. (1993). The nature of dispute and the effectiveness of international mediation. *Journal of Conflict Resolution, 37*(4), 670–691.

Bercovitch, J., & Wells, R. (1993). Evaluating mediation strategies - a theoretical and empirical analysis. *Peace & Change, 18*(1), 3–25.

Bond, D., Bennet, B., & Voegele, W. B. (1994). Panda: Interaction events data development using automated human coding.. Extended Version of a paper presented at the 1994 Annual Meeting of the International Studies Association in Washington, DC on April 1st, 1994.

Brecher, M., Wilkenfeld, J., & Moser, S. (1988). *Crises in the Twentieth Century - Handbook of International Crises*, Vol. I. Pergamon Press, Oxford.

Butterworth, R. L. (1976). *Managing Interstate Conflict, 1945-74: Data with Synopses*. University of Pittsburgh Center for International Studies, Pittsburgh.

Fürnkranz, J., Petrak, J., & Trappl, R. (1997). Knowledge discovery in international conflict databases. *Applied Artificial Intelligence, 11*(2), 91–118.

Gantzel, K. J., & Meyer-Stamer, J. (Eds.). (1986). *Die Kriege nach dem zweiten Weltkrieg bis 1984*. München.

Gochman, C. S., & Maoz, Z. (1984). Militarized interstate disputes 1816–1976: Procedures, patterns, and insights. *Journal of Conflict Resolution, 28*, 585–616.

Holsti, K. J. (1983). *International Politics: A Framework for Analysis* (2nd edition). Englewood Cliffs.

Hudson, V. M. (1991). *Artificial Intelligence and International Politics*. Westview Press, Boulder, CO.

John, G. H., Kohavi, R., & Pfleger, K. (1994). Irrelevant features and the subset selection problem. In Cohen, W., & Hirsh, H. (Eds.), *Proceedings of the 11th International Conference on Machine Learning (ML-94)*, pp. 121–129 New Brunswick, NJ. Morgan Kaufmann.

Kohavi, R., John, G., Manley, D., & Pfleger, K. (1994). MLC++: A machine learning library in C++.. In *Tools with Artificial Intelligence*. IEEE Computer Society Press.

Kononenko, I., & Bratko, I. (1991). Information-based evaluation criterion for classifier's performance. *Machine Learning, 6*, 67–80.

Mallery, J. C. (1988). Thinking about foreign policy: Finding an appropriate role for artificial intelligence computers. Master's thesis, M.I.T. Political Science Department, Cambridge, MA.

Mallery, J. C., & Sherman, F. L. (1993). Learning historical rules of major power intervention in the post-war international system.. Paper prepared for presentation at the 1993 Annual Meeting of the International Studies Association.

Petrak, J. (1994). VIE-CBR - Vienna case-based reasoning tool, version 1.0: Programmer's and installation manual. Technical report TR-94-34, Austrian Research Institute for Artificial Intelligence, Vienna. Available from ftp://ftp.ai.univie.ac.at/papers/oefai-tr-94-34.ps.Z.

Petrak, J., Trappl, R., & Fürnkranz, J. (1994). The possible contribution of AI to the avoidance of crises and wars: Using CBR methods with the KOSIMO database of conflicts. Technical report TR-94-32, Austrian Research Institute for Artificial Intelligence, Vienna. Available from ftp://ftp.ai.univie.ac.at/papers/oefai-tr-94-32.ps.Z.

Pfetsch, F. R., & Billing, P. (1994). *Datenhandbuch nationaler und internationaler Konflikte*. Nomos Verlagsgesellschaft, Baden-Baden.

Quinlan, J. R. (1993). *C4.5: Programs for Machine Learning*. Morgan Kaufmann, San Mateo, CA.

Schrodt, P. A. (1991a). Artificial intelligence and international relations: An overview. In (Hudson, 1991).

Schrodt, P. A. (1991b). Classification of interstate conflict outcomes using a bootstrapped ID3 algorithm. *Political Analysis*.

Schrodt, P. A. (1996). *Patterns, Rules and Learning: Computational Models of International Behavior*. University of Michigan Press. Forthcoming.

Schrodt, P. A., & Davis, S. G. (1994). Techniques and troubles in the machine coding of international event data. Dept. of Political Science, University of Kansas. Paper presented at the 1994 meeting of the International Studies Association, Washington DC.

Sherman, F. L. (1988). Sherfacs: A new cross-paradigm, international conflict dataset.. Paper written for presentation at the 1988 annual meeting of the International Studies Association.

Trappl, R. (1986). Reducing international tension through artificial intelligence: A proposal for 3 projects. In Trappl, R. (Ed.), *Power, Autonomy, Utopia: New Approaches Toward Complex Systems*. Plenum, New York.

Trappl, R. (1992). The role of artificial intelligence in the avoidance of war. In Trappl, R. (Ed.), *Cybernetics and Systems '92*, pp. 1667–1672 Singapore. World Scientific.

Unseld, S. D., & Mallery, J. C. (1993). Interaction detection in complex datamodels.. MIT A.I. Memo No. 1298.

Vogele, W. B. (1994). Global conflict profiles: Event data analysis using panda.. Paper prepared for presentation at the annual meeting of the American Political Science Association, New York City, Sep. 1994.

Wilkenfeld, J., Brecher, M., & Moser, S. (1988). *Crises in the Twentieth Century - Handbook of Foreign Policy Crises*, Vol. II. Pergamon Press, Oxford.

A   Pure Rules discovered in the CONFMAN database

Rules contained in an unpruned decision tree that cover 10 or more *unsuccessful* conflict management attempts:

```
Rule F1:
if (400 < V5_FAT <= 700000) &&
   (V20_POWERB <= 33) &&
   (V32_LIBA <= 3) &&
   (V33_LIBB == 1) &&
   (V35_MGMTACT == "MEDIATION")
then SUCCESS: 0
     FAILURE: 12 (2 Conflicts)
```

```
Rule F2:
if (400 < V5_FAT <= 700000) &&
   (V20_POWERB <= 33) &&
   (V24_POLSYSA == "MULTI-PARTY") &&
   (V32_LIBA <= 3) &&
   (V33_LIBB > 1) &&
   (V35_MGMTACT == "MEDIATION") &&
   (39 < V43_TIM <= 67)
then SUCCESS: 0
     FAILURE: 19 (4 Conflicts)
```

```
Rule F3:
if (V3_DUR > 6) &&
   (400 < V5_FAT <= 700000) &&
   (V20_POWERB <= 33) &&
   (V24_POLSYSA == "MULTI-PARTY") &&
   (V32_LIBA <= 3) &&
   (V33_LIBB > 1) &&
   (V35_MGMTACT == "MEDIATION") &&
   (V40_NRMED > 3) &&
   (67 < V43_TIM <= 256)
then SUCCESS: 0
     FAILURE: 14 (3 Conflicts)
```

```
Rule F4:
if (V3_DUR > 76) &&
   (400 < V5_FAT <= 700000) &&
   (V19_POWERA <= 31) &&
   (V20_POWERB <= 22) &&
   (V30_RIGHTSA > 4) &&
   (V32_LIBA > 3) &&
   (V35_MGMTACT == "MEDIATION") &&
   (V38_MEDSTR == "DIRECTIVE") &&
   (V41_NRMEDM <= 4) &&
   (61 < V43_TIM <= 136)
then SUCCESS: 0
     FAILURE: 16 (5 Conflicts)
```

```
Rule F5:
if (V3_DUR <= 190) &&
   (400 < V5_FAT <= 700000) &&
   (V19_POWERA <= 31) &&
   (V20_POWERB <= 33) &&
   (V32_LIBA > 3) &&
   (V33_LIBB > 3) &&
   (V35_MGMTACT == "MEDIATION") &&
   (V38_MEDSTR == "COMM-FACIL") &&
   (V39_RELMED == "NO_PREV_REL") &&
   (V41_NRMEDM <= 2) &&
   (V93_HOMT == "MAJORITY")
then SUCCESS: 0
   FAILURE: 14 (8 Conflicts)
```

```
Rule F6:
if (400 < V5_FAT <= 700000) &&
   (V19_POWERA > 31) &&
   (V20_POWERB <= 33) &&
   (V32_LIBA > 3) &&
   (V35_MGMTACT == "MEDIATION") &&
   (V41_NRMEDM <= 4) &&
   (V43_TIM <= 158)
then SUCCESS: 0
     FAILURE: 13 (3 Conflicts)
```

```
Rule F7:
if (V5_FAT > 700000) &&
   (V43_TIM <= 73) &&
   (V84_TISC == "DIFF_TIME_SYS")
then SUCCESS: 0
     FAILURE: 31 (3 Conflicts)
```

Rules contained in an unpruned decision tree that cover 10 or more *successful* conflict management attempts:

```
Rule S1:                              Rule S2:
if (1 < V3_DUR <= 3) &&               if (V2_DUR_G > 2) &&
   (V5_FAT <= 400) &&                    (V5_FAT <= 400) &&
   (V20_POWERB <= 33) &&                 (V20_POWERB <= 33) &&
   (V35_MGMTACT == "MEDIATION")          (V35_MGMTACT == "MEDIATION") &&
then SUCCESS: 15                         (V43_TIM <= 35) &&
     FAILURE: 0 (8 Conflicts)            (V44_REQINI == "BOTH_PARTIES")
                                      then SUCCESS: 15
                                           FAILURE: 0 (11 Conflicts)


Rule S3:                              Rule S4:
if (400 < V5_FAT <= 700000) &&        if (V3_DUR > 13) &&
   (V19_POWERA <= 31) &&                 (400 < V5_FAT <= 700000) &&
   (V20_POWERB <= 22) &&                 (V19_POWERA <= 31) &&
   (V32_LIBA > 3) &&                     (V20_POWERB <= 33) &&
   (V35_MGMTACT == "MEDIATION") &&       (V32_LIBA > 3) &&
   (V38_MEDSTR == "DIRECTIVE") &&        (V35_MGMTACT == "MEDIATION") &&
   (V41_NRMEDM <= 4) &&                  (V38_MEDSTR == "PROCEDURAL") &&
   (V43_TIM > 145) &&                    (V43_TIM <= 18) &&
   (V84_TISC == "DIFF_TIME_SYS")         (V94_TIS <= 4)
then SUCCESS: 14                      then SUCCESS: 10
     FAILURE: 0 (5 Conflicts)             FAILURE: 0 (5 Conflicts)


Rule S5:
if (815 < V5_FAT <= 700000) &&
   (V19_POWERA <= 31) &&
   (V20_POWERB <= 10) &&
   (V30_RIGHTSA > 4) &&
   (V32_LIBA > 3) &&
   (V35_MGMTACT == "MEDIATION") &&
   (V38_MEDSTR == "DIRECTIVE") &&
   (V39_RELMED == "SAME_BLOC_BOTH") &&
   (V40_NRMED <= 5) &&
   (V41_NRMEDM <= 2) &&
   (V43_TIM <= 61)
then SUCCESS: 12
     FAILURE: 0 (7 Conflicts)
```

LIST OF PARTICIPANTS

LIST OF PARTICIPANTS

Prof. Dr. Hans-Hermann Bock
RWTH Aachen
Institute fuer Statistik
Wuellnerstrasse 3
D-52056 Aachen (GERMANY)

Prof. Ivan Bratko
J. Stefan Institue
Jamova 39
1000 Ljubljana (SLOVENIA)

Dr. Hans Hellendoorn
Siemens AG
ZFE T SN 4
(Otto-Hahn Ring 6)
D-81730 Muenchen (GERMANY)

Prof. Yves Kodratoff
Université de Paris-Sud
Centre d'Orsay
Lab. de Recherche en Informatique
Bâtiment 490
F-91405 Orsay Cedex (FRANCE)

Prof. Dr. Rudolf Kruse
University of Magdeburg
Universitaetsplatz 2
D-39106 Magdeburg (GERMANY)

Prof. Nada Lavrac
J. Stefan Institue
Jamova 39
1000 Ljubljana (SLOVENIA)

Prof. Dr. Hans-Joachim Lenz
Freie Universität Berlin
Institut für Statistik und Ökonometrie
Garystr. 21
D-14195 Berlin (GERMANY)

Dr. Rolf A. Müller
Daimler-Benz AG
Forschung und Technik
Alt-Moabit 96a
D-559 Berlin (GERMANY)

Prof. Sarunas Raudys
Inst. of Mathematics and Informatics
Akademijos 4
Vilnius 2600 (LITHUANIA)

Prof. Dr. Helge Ritter
Universität Bielefeld
Technische Fakultät
AG Neuroinformatik
Postfach 10 01 31
D-33501 Bielefeld (GERMANY)

Prof. Dr. Egmar Rödel
Institut für Informatik
Humboldt Univ. zu Berlin
Unter den Linden 6
D-10099 Berlin (GERMANY)

Prof. Lorenza Saitta
University of Torino
Dipartimento di Informatica
Corso Svizzera 185
I-10149 Torino (ITALY)

Prof. Dr. Robert Trappl
Austrian Research Institute
for Artificial Intelligence
Schottengasse 3
A-1010 Wien (AUSTRIA)

Prof. Dr. Fritz Wysotzki
Fachbereich 13 Informatik
Technische Universität Berlin
Franklinstr. 28/29
D-10587 Berlin (GERMANY)

Dr. Giorgio Brajnik
Dip. di Matematica e Informatica
University of Udine
via delle Scienze, 206
I-33100 Udine (ITALY)

Dr. Marina Campolo
Dip. Scienze e Tecnologie Chimiche
University of Udine
via del Cotonificio, 108
I-33100 Udine (ITALY)

Prof. Giacomo Della Riccia
Dip. di Matematica e Informatica
University of Udine
via delle Scienze, 206
I-33100 Udine (ITALY)

Dr. Francesco Fabris
Dip. di Matematica e Informatica
University of Udine
via delle Scienze, 206
I-33100 Udine (ITALY)

Dr. Gian Luca Foresti
Dip. di Matematica e Informatica
University of Udine
via delle Scienze, 206
I-33100 Udine (ITALY)

Dr. Andrea Fusiello
Dip. di Matematica e Informatica
University of Udine
via delle Scienze, 206
I-33100 Udine (ITALY)

Dr. Stefano Mizzaro
Dip. di Matematica e Informatica
University of Udine
via delle Scienze, 206
I-33100 Udine (ITALY)

Dr. Vittorio Murino
Dip. di Matematica e Informatica
University of Udine
via delle Scienze, 206
I-33100 Udine (ITALY)

Dr. Stefano Soatto
Dip. di Matematica e Informatica
University of Udine
via delle Scienze, 206
I-33100 Udine (ITALY)

Prof. Maria Staniszkis
Dip. di Matematica e Informatica
University of Udine
via delle Scienze, 206
I-33100 Udine (ITALY)

Printed in the United States
By Bookmasters
Printed in the United States
By Bookmasters